Peripheral Visions

Peripheral Visions

LEARNING
ALONG THE WAY

MARY CATHERINE BATESON

■ HarperCollins*Publishers*

HarperCollins books may be purchased for educational, business, or sales promotional use. For information, please write: Special Markets Department, HarperCollins Publishers, Inc., 10 East 53rd Street, New York, NY 10022.

FIRST EDITION

Designed by George J. McKeon

Library of Congress Cataloging-in-Publication Data

Bateson, Mary Catherine.
    Peripheral visions : learning along the way / Mary Catherine Bateson.
        p.   cm.
    Includes bibliographical references.
    ISBN 0-06-016859-5
    1. Experiential learning. 2. Autobiographical memory. 3. Bateson, Mary Catherine. I. Title.
    BF318.5B38   1994
    001.2—dc20   93-45983

94 95 96 97 98 ❖/RRD 10 9 8 7 6 5 4 3 2 1

*Until you are at home somewhere, you cannot be at home everywhere.**

For my family, that home, in learning and return, has been the Monadnock region of New Hampshire. This book, which speaks so often of distant places, is dedicated in celebration of three decades of friendship with Rose Mary Eldredge and Mary Garland and the families they have nurtured there.

* With apologies to Howard Thurman.

Contents

1. Improvisation in a Persian Garden   1

2. Learning from Strangers   15

3. Double Helix   29

4. Something Blue   45

5. A Mutable Self   59

6. Construing Continuity   77

7. Attending a World   95

8. Longitudinal Epiphanies   111

9. Turning into a Toad   127

10. Joining In   144

11. Composing Our Differences   161

12. Limited Good   179

13. Learning as Coming Home   195

14. The Seemly and the Comely   214

15. Reflected Visions   227

Acknowledgments and Sources   237

I

Improvisation in a Persian Garden

WE BEGIN IN A PERSIAN GARDEN. In summer it would be filled with scent and color, but on this midwinter day, some twenty years ago, only the bones of the garden and its geometries were visible: a classic walled garden, with ranks of leafless fruit trees between patches of dirty snow and rosebushes recognizable by their thorns. There was a row of dark cypresses along one end and a dry watercourse down the center, clogged with leaves.

I had arrived in Tehran in January 1972 with my husband, Barkev, and our two-and-a-half-year-old daughter, Sevanne, known as Vanni, for the beginning of a period of research, teaching, and institution building in a culture and language new to both of us, even though I had studied the Islamic tradition in the Arab world and Barkev had grown up as an Armenian Christian in Aleppo, Syria. The day before, we had gone to tea—in fact, cocktails—at the home of our new landlords, and when they heard that I was an anthropologist, they invited us to come with them the next day to a garden on family land in a village near Tehran, where they would observe Eyd-e Qorban, the Feast of Sacrifice.

At the same time that pilgrims are performing the Meccan pilgrimage, Muslims around the world are celebrating some of the steps of that weeklong ritual, one of which is an animal sacrifice, usually a sheep or a camel. It is said that in Saudi Arabia so many sheep and camels are slaughtered by pilgrims that they are simply plowed under by bulldozers. Many families in Iran also sacrifice a sheep on the Feast of Sacrifice, and here the meat is traditionally given to the poor. When our landlady invited us, although Barkev could not come, well, of course I said I would, but I would have to bring Vanni with me.

On the way to the country, I began to have second thoughts and found myself running a mental race with the car, trying to work out the implications of the invitation I had lightheartedly accepted: to come and see a sheep being slaughtered, bringing a two-year-old child. I worried about what to say to her, how she would react. I went back over the memory of seeing the necks of chickens wrung when I was a child, before chickens came reliably headless, neatly butchered, and wrapped in plastic on little nonbiodegradable white trays. Those who eat meat, I told myself, should at least know where it comes from. That trip to Iran was not the first time I had entered a strange culture, but it was the first time I had done so with a child, and this was the first of many moments when the double identity of mother and field-worker led me down new paths of reflection.

I know now that to be in such a garden was to stand in the middle of a vision of the world. In the Persian tradition, a garden is itself a cosmological statement, a diagram reiterated in the design of carpets and the way they are used. Gardens are generally symmetrical in plan, but within that ordered framework the rich particularity is varied and relaxed. Gardens are bounded, walled; within, all is fertile and hospitable, but there is always an awareness of a world outside that is less benign, an unruly and formless realm of desert harshness and marauding strangers.

Water is a part of every garden, either flowing water or a pool large enough for the ritual washing that precedes prayer. House agents used to point out the pools to Westerners as wading pools for their children, but in fact these pools, reflecting and doubling garden and sky, focus the ideals of purity and generosity. The shah built a palace in the northern part of Tehran, at the high edge of the city, so water from the palace garden would symbolically flow down through the city to his subjects. Every garden is a promise of the paradise to which the faithful will go after death.

It was a cold, gray day, and all of us were bundled up. My hostess, elegant in a fur coat and stylish boots, explained that the butchering would be done by the gardener. She herself, she said, could never bear to look on at the moment the sheep was killed. When we arrived we found the gardener and his whole family waiting: his wife, wearing the dark printed cotton veil of a village woman, and three children of different ages. Our gathering modeled the tensions of Iranian society before the revolution: the affluent, Westernized urbanites, the villagers, the performance of a ritual that rooted all of them in the past—all this in a setting with its own affirmations about the relationships between God and humankind and the nature of the cosmos.

The sacrifice of a sheep links several religious traditions. The Patriarch Abraham, we are told, was commanded by God to sacrifice his son—Ishmael in the Islamic version, Isaac for Jews and Christians; then, at the last moment, he was provided with a ram as a substitute. The story is emphasized most in Islam, representing the value of total submission to the will of God, a God who proves compassionate. Volumes have been written about the relationship between this single observance and older traditions of human and animal sacrifice and their echoes in Jewish, Christian, and Islamic worship. In the rituals of Passover and Easter, the recurrent sacrifice is reshaped in various ways: the shank bone of a lamb on the Seder table, the Lamb of God imaged in the single atoning sacrifice of Good Friday and its year-round memorials. There in that garden an ancient symbol connecting many faiths was still a real sheep that would bleed real blood.

The sheep was given a drink of water and turned toward the south, toward Mecca. Then, saying "in the name of God" and "Allahu Akbar," the gardener quickly slit its throat, letting the blood gush out into a ditch. I was holding Vanni on my hip and explaining what was happening, for children can handle such scenes (including those we see every day on television) if they see them in the company of an adult who both interprets and sets an emotional tone.

The sheep was skinned and its various organs removed. The gardener made a slit above one hind hoof and began to blow into it, forcing air under the skin to separate it from the flesh, so the body gradually puffed out and he could slit down the front and remove the woolly pelt. Deftly he spread the fleece, wool side down, on the ground, to receive the various edible organs one by one. "See, Vanni," I said, "that's a heart; every animal has one. Its job is pushing the blood around the body all the time." I wondered if I should introduce the word *pump*? No time. Intestines, stomach, liver . . . some were harder than others to explain, and my mind was racing for vocabulary to convey an understanding of the similarity of all mammals that would not trigger too much identification with the dead sheep. The fleece and the organs neatly laid out upon it would be the butcher's portion.

I had slipped into a teaching role, taking advantage of the visual aids to give a minilecture about each organ, using a vocabulary appropriate to a verbal two-and-a-half-year-old. Then suddenly, just as I was saying, "That big thing is a lung, see, for pulling in air and breathing, and the sheep has two of them, just like we do—here comes the other," I thought, with a moment of intense shock, Why, it's huge. In fact, I had never observed the death or dissection of any mammal larger than a mouse. Because I had an abstract knowledge of anatomy, distanced from the reality, and because I was preoccupied with Vanni's experience, it was almost impossible to realize that I was encountering something new myself. What I was passing on to her was not knowledge based on direct experience but a set of labels, whose theoretical character was invisible to me until I was jolted by a detail I had failed to anticipate. My words could hardly have been more abstract if I had started from theology or Biblical history.

Just as I could say, "That's a heart, that's a lung," we go through life, saying, "I must be in love," "Oh, this is seasickness," "This is an orgasm," "This is a midlife crisis." We are ready with culturally constructed labels long before we encounter the realities, even to the point of saying, "This is a heart attack," "I must be dying." We can call our fate by name before we meet it. It will not retreat, but we are often relieved when doctors name our conditions.

The facts of the body both separate and connect. They testify to the links between human beings and other mammals and living systems, but they divide the sexes and the developmental stages. The body's truths are often concealed, so it is not always easy to learn about birth or sex or death, or the curious and paradoxical relationships between them. We keep them separate and learn about them on different tracks, just as we learn separately about economics and medicine and art, and only peripheral vision brings them back together. Experience is structured in advance by stereotypes and idealizations, blurred by caricatures and diagrams.

In the past, we might have climbed as children onto a big four-poster used in turn for birthing, lovemaking, and dying. In other cultures we might have grown up seeing human bodies of every age: torsos as distinctive as faces; breasts of many forms and stages, from the barely budding to the flat, long breasts of multiple lactations; penises as varied as noses. Living in a palm and bamboo hut or under a roof without surrounding walls, we would have listened at night to the sounds people make when their bodies are busy with the primordial efforts of pain or pleasure. We might have seen the interiors of human bodies as well, in places where custom demands the dismemberment of the newly dead. In the modern West, however, even intimacy is categorized and filtered through abstractions.

Anthropologists are trained to be participants and observers at the same time, but the balance fluctuates. Sometimes a dissonance will break through and pull you into intense involvement in an experience you had distanced by thinking of yourself as coolly looking on. Or it may push you away when you have begun to feel truly a part of what is happening. I was in that garden as a learner, an outsider, and yet, because I was there as a parent, I was simultaneously a teacher, an authority. Trying to understand and remember what I saw, I was also trying to establish an interpretation that would be appropriate for Vanni, one that would increase her understanding of the living world and her place in it and also bring her closer to the Iranians she would be living among for several years. At least, I wanted to leave her unfrightened. Out of that tense multiplicity of vision came the possibility for insight.

That day in the Persian garden has come to represent for me a changed awareness of learning pervading other activities. Meeting as strangers, we join in common occasions, making up our multiple roles as we go along—young and old, male and female, teacher and parent and lover—with all of science and history present in shadow form, partly illuminating and partly obscuring what is there to be learned. Mostly we are unaware of creating anything new, yet both perception and action are necessarily creative. Much of modern life is organized to avoid the awareness of the fine threads of novelty connecting learned behaviors with acknowledged spontaneity. We are largely unaware of speaking, as we all do, sentences never spoken before, unaware of choreographing the acts of dressing and sitting and entering a room as depictions of self, of resculpting memory into an appropriate past.

This awareness is newly necessary today. Men and women confronting change are never fully prepared for the demands of the moment, but they are strengthened to meet uncertainty if they can claim a history of improvisation and a habit of reflection. Sometimes the encounter takes place on journeys and distant sojourns, as it has for me in periods of living in Israel, the Philippines, and Iran. Often enough we encounter the strange on familiar territory, midway through familiar actions and commitments, as did the Iranian gardener whose cosmopolitan employers had become half foreign to him. Sometimes change is directly visible, but sometimes it is apparent only to peripheral vision, altering the meaning of the foreground.

What I tried to do that day, stringing together elements of previous knowledge, attending to catch every possible cue, and exploring different translations of the familiar, was to improvise responsibly and with love. Newly arrived in Iran, I had no way of knowing what was going to happen, not even a clear sense of my own ignorance. Even so, I was trying to put together a way of acting toward my child and my hosts that would allow all of us, in courtesy and goodwill, to sustain a joint performance.

Vanni of course was generating a novel performance too, trying to figure out who to be and how to react, the complex perennial task of childhood. She got some of her cues from me, but she also kept a watchful eye on the children of the gardener: the oldest, who watched as he had many times before, with a sense of occasion and none of horror, and the younger ones gleaning the confidence that this was an ordinary, unfrightening process taking place in front of them, but a solemn and even festive moment as well, one that would be repeated and explored in play.

So there we were, nine people differing in at least four dimensions: adults and children, females and males, Iranians and Americans, affluent urbanites and villagers, with differences of language and religion falling along the same cleavages. We were joined in the performance of a ritual, in spite of the fact that we did not share a common script or common doctrines. What was happening had different meanings to each of us. The contrasts were as great between the sophisticated urban people and the villagers, who were all nominally Muslims and Iranians, as between the American outsiders and our hosts. Men and women, nominally sharing the same culture, must bridge comparable gaps, yet for better or worse they have always done so, for all human beings live with strangers.

Occasions like this encounter in a winter garden provide the frameworks for future learning. Both on the scene and in memory, similarities are tasted and compared. Vanni was learning something about my stance that would affect her play with neighbor children in the alley behind our house. By now she has forgotten that day, but she would have remembered it over the next few months, and other experiences would have been matched with it and sorted out in her learning. She was going through the process of immersion in a second culture early in her learning of a first and having to adapt not only to what adults can explain but also to things for which they have no words. Children cope superbly where anthropologists must grope. I believe that participant observation is more than a research methodology. It is a way of

being, especially suited to a world of change. A society of
many traditions and cultures can be a school of life. Even as
a two-year-old in that scene, Vanni had to improvise not one
but multiple roles.

The quality of improvisation characterizes more and
more lives today, lived in uncertainty, full of the inklings of
alternatives. In a rapidly changing and interdependent world,
single models are less likely to be viable and plans more
likely to go awry. The effort to combine multiple models risks
the disasters of conflict and runaway misunderstanding, but
the effort to adhere blindly to some traditional model for a
life risks disaster not only for the person who follows it but
for the entire system in which he or she is embedded, indeed
for all the other living systems with which that life is linked.

Adaptation comes out of encounters with novelty that
may seem chaotic. In trying to adapt, we may need to deviate
from cherished values, behaving in ways we have barely
glimpsed, seizing on fragmentary clues. The improvisatory
artist cannot be sure whether a given improvisation will
stand as a work of art or be rejected as an aberration.
Trusted habits of attention and perception may be acting as
blinders. Resources we have relied on to shape our lives may
turn out to be dangerous addictions or spin into new shapes
as the earliest versions of emerging patterns. Essential
themes are not clearly marked but rather visible only out of
the corner of the eye.

Under the pressure of the moment, needing to respond, it
is easy to be captured by some central point of focus. A dead
sheep. The spilling of blood and its impact on a small child.
But there is always more in any episode, much of it at the
very edge of awareness, most of it in flux, the relationships
within any cultural tradition between old and new barely visible.

This same ambiguity sets new tasks for parents and
teachers. Instead of passing on hallowed certainties and
maintaining the status quo, they must make childhood an
open-ended introduction to a process of continual change in
which self-observation can become the best of teachers. If we
knew the future of a particular child, we might be able to
prepare that child with all the necessary skills and attitudes,

and we might say at a given moment that the preparation is completed and it is time for real life to commence. That situation, however, is long gone, if indeed it ever existed. Rarely is it possible to study all the instructions to a game before beginning to play, or to memorize the manual before turning on the computer. The excitement of improvisation lies not only in the risk involved but in the new ideas, as heady as the adrenaline of performance, that seem to come from nowhere. When the necessary tasks of learning cannot be completed in a portion of the life cycle set aside for them, they have to join life's other tasks and be done concurrently. We can carry on the process of learning in everything we do, like a mother balancing her child on one hip as she goes about her work with the other hand or uses it to open the doors of the unknown. Living and learning, we become ambidextrous.

Systems of education are everywhere in ferment, visions of promise countered with proposals for increasing rigidity. The Japanese are as puzzled as the Americans—ostensibly for opposite reasons—and you can find editorials about the flaws of education systems from London to Manila. This suggests that many proposals have too narrow a focus, are directed at local problems when the entire concept of education needs to be rethought. Looking at the place of learning in other societies and times from this vantage point, it is reassuring to know that everywhere most of learning occurs outside the settings labeled as educational. Living and learning are everywhere founded on an improvisational base. The discovery of new needs may be followed by adding units to the syllabus, but it can also lead to the discovery of how human beings make do with partial understandings, invent themselves as they go along, and combine in complex undertakings without full agreement about what they are doing. These skills also are learned.

Ambiguity is the warp of life, not something to be eliminated. Learning to savor the vertigo of doing without answers or making shift and making do with fragmentary ones opens up the pleasures of recognizing and playing with pattern, finding coherence within complexity, sharing within multiplicity. Improvisation and new learning are not private

processes; they are shared with others at every age. The multiple layers of attention involved cannot safely be brushed
aside or subordinated to the completion of tasks. We are
called to join in a dance whose steps must be learned along
the way, so it is important to attend and respond. Even in
uncertainty, we are responsible for our steps.

Starting from a Persian garden, this discussion will wander the paths of attention and improvisation in and out of
four countries and across the life cycle. We will look for the
sources of the bits and pieces stitched into improvisations and
for the underlying stiffener that unifies; for the habits of learning and ways of building a repertoire from which to improvise, the metaphors that link one experience to another.
Shared ways of seeing are socially constructed and currently,
fashionably, criticized and deconstructed, but when you are
able to attend to something new or to see the familiar in a new
way, this is a creative act. I would call it a godlike act, except
that the word evokes, for too many, a sense of distance and
dominance, while seeing anew is a kind of intimacy; I would
call it childlike, if it were not important to avoid blocking
learning with the reminders of all that was onerous in childhood. In the ordinary creativity of moving through the world,
we are both gods and children.

Increasingly, we will cease to focus on learning as prelim-
inary and see it threaded through other layers of experience,
offering one of life's great pleasures. There is this truth in
reinforcement theory, that pleasure and survival are linked
by learning. The capacity to enjoy, to value one experience
over another, is the precondition of the capacity to learn, so
that even the innate sucking reflex of infants is shaped
within hours of birth by the rewards of success. Because
learning is the most basic of human adaptive processes, we
can hope that it will lead toward a relationship with the rest
of the biosphere that is both satisfying and sustainable.

There is a spiritual basis to attention, a humility in waiting upon the emergence of pattern from experience. The willingness to assimilate what has been seen or heard draws
other life into increasingly inclusive definitions of the self.
Looking, listening, and learning offer the modern equivalent

of moving through life as a pilgrimage. Even death is a time to learn.

This is a book of stories and reflections strung together to suggest a style of learning from experience. Wherever a story comes from, whether it is a familiar myth or a private memory, the retelling exemplifies the making of a connection from one pattern to another: a potential translation in which narrative becomes parable and the once upon a time comes to stand for some renascent truth. This approach applies to all the incident of everyday life: the phrase in the newspaper, the endearing or infuriating game of a toddler, the misunderstanding at the office. Our species thinks in metaphors and learns through stories.

Many tales have more than one meaning. It is important not to reduce understanding to some narrow focus, sacrificing multiplicity to what might be called the rhetoric of merely: merely a dead sheep, only an atavistic ritual, nothing but a metaphor. Openness to peripheral vision depends on rejecting such reductionism and rejecting with it the belief that questions of meaning have unitary answers. Twenty years after it occurred, in a world increasingly troubled by ethnic conflict, a remembered ritual on the Feast of Sacrifice came to exemplify for me layer upon layer of processes whereby human beings can join and communicate and learn in spite of profound differences. The story grew into this book as memories from Iran resonated with memories of years lived in Israel and in the Philippines. A Persian garden has become a magic carpet. The process of spiraling through memory to weave connection out of incident is basic to learning, so that in this and perhaps other ways the text is a demonstration of its subject matter.

In my work, I have always been able to start from listening and looking, doing natural history along with the myriad small experiments that occur when one tries, tentatively, deferentially, to join in. These are skills that spill over into all areas of life. I cannot know which observation will propose a theme that proves key to understanding, narrowing in attention to a pattern that opens out to many others. It is common to gather data in fieldwork and continue to mine that data years later to illuminate questions still unposed when the original material was collected. Records of the past or of strange peoples in far places tell us not what always happens but what can happen, the depth of the human potential, the range of the capacity to muddle through. No single narrative is sufficient for understanding, no single model sufficient for aspiration. Even though they are spread around the globe, the societies—Jewish, Christian, and Islamic—that I draw on for examples still have much in common, combining elements of Semitic and Greek traditions in their attitudes, still all acting in the shadow of Abraham. The human range is both wider and deeper.

My mother used to list paths to "insight" in her lectures: anthropological fieldwork, the study of infants, the study of another species, psychoanalysis, the experience of either a psychosis or a religious conversion (followed by recovery), or "a love affair with an Old Russian." The stories in this book do not cover all that ground, but the list is not a bad road map. Sometimes a narrative which seems to fit into one category metamorphoses into another. These are all ways of learning, by encountering and comparing more than one version of experience, that the realities of self and world are relative, dependent on context and point of view. Because we live in a world of change and diversity, we are privileged to enter, if only peripherally, into a diversity of visions, and beyond that to include them in the range of responsible caring.

We live not only in the presence of different cultural visions but with different individual modes of perception, with access to the memories of childhood and of alternative states of consciousness. These resonate with the many layers of vision within any single cultural tradition, the mythic and the multiply metaphorical, the sacred and the invisibly empirical, the insights of the laboratory and those of poetry and sleep. To become open to multiple layers of vision is to be both practical and empathic, to practice the presence of God or gods and to practice wilderness. Learning the paths of human culture, we are attentive as well to the undomesticated outdoors and the essential wildness spinning on in subatomic spaces, forever generating new patterns.

Courtesy is one of the great human inventions for bridging uncertainty. On the day of that ambiguous celebration of the Feast of Sacrifice, the gardener's wife brought glasses of tea as the shadows lengthened. Our landlords packed the meat in plastic to take to Tehran, where it would be frozen to share with family members. We got in the car and rode back into the city, with Vanni falling asleep on my lap. We were in Iran, increasingly at home, for most of the next seven years, until the revolution.

By now, that wealthy urban couple may be outside the country, learning to live in a new place. One of the gardener's sons may have grown up to die in the war with Iraq or been drafted to meet other new challenges, while the daughters have been held in more narrow roles. Everyone's life has been redirected by political change. Yet the revolution that occurred seven years later was present in that garden, present in the class differences and the ambivalent relationship to tradition of the wealthy urban landowners, the cultural disparities between them and the villagers they were employing, the suspect presence of a foreigner. Iran was a society in which ways of understanding became unmanageably disparate, for the rise of fundamentalism within any tradition is always a symptom of the unwillingness to try to sustain joint performances across disparate codes—or, to put it differently, to live in ambiguity, a life that requires constant learning. The risk of such a failure is the challenge that faces our society—our entire species—today.

That brief encounter in a Persian garden offered its participants many kinds of experience. There was room for hostility and anxiety, for fear of strangeness and distaste at reminders of the flesh and of mortality. There was room for awe in the presence of one of humankind's transcendent visions imposing its abstract geometries, for in Islam all space is ordered in relation to Mecca and all time in relation to salvation and judgment day. The sacred was represented and so was the organic, intimacy and strangeness. There was room for boredom and for embarrassment and awkwardness. With so many layers of possibility, there was room for a great deal of learning, but reason too for rejecting learning. I remember my fingers and toes growing cold.

In any experience other moments are present, and so they are here. We will move within paragraphs from a Philippine village to the Sinai desert. When we are in a Persian garden we will be at the same time in my New Hampshire studio. When I write about Vanni at age two I am accompanied by a teenager and by a woman in her twenties. Because the presence of many kinds of attention is central to my theme, I have included these other layers of awareness in my text; they are all relevant. *Insight*, I believe, refers to that depth of understanding that comes by setting experiences, yours and mine, familiar and exotic, new and old, side by side, learning by letting them speak to one another.

2

Learning from Strangers

OVER THE YEARS, Vanni has often given me "desk friends," small, friendly items grouped behind the computer to greet me during the day, familiar spirits. A small globe, blue and green. A beanbag salamander. The latest is an ammonite, the fossil of a prehistoric mollusk like a chambered nautilus. This graceful spiral of growth is related to the "golden section" of aesthetics and to the mathematical pattern called a Fibonacci series, in which each number is equal to the sum of the two preceding, increasing rapidly. *Cornu Ammonis*, the horn of Amen, Egyptian god of life and reproduction, represented as a man with a ram's head. Ideas are coiled one into another, the Mesozoic decoded in the Renaissance and architecture meeting with biology in what H. E. Huntley calls "mathematical beauty."

These tokens came with me to the MacDowell Colony, the artists' colony in Peterborough, New Hampshire, where much of this book was written, along with reminders of the countries where I have worked. Each of us—writers, painters, composers spending a month or two in residence—had a studio in the woods for pursuing separate visions, where no one came unless invited. The studios of writers are the smallest, needing little but a worktable, but I chose what I brought carefully, spaced out where it would catch my eye to evoke the memories I was seeking to connect.

Because I would be writing about Iran, I even brought a small tribal carpet, made to be loaded and unloaded by nomads pitching their tents in new grazing grounds from day to day, suitable for temporary habitation. I hung a Philippine Christmas ornament in my studio window, a mandala of palm fronds, the pervasive material of village ornament and construction; on my desk I put a small, carved figure of Christ, one of hundreds detached from crucifixes and sold in Manila as antiquities, usually armless, with their pointed European noses sliced off during the era when Filipinos rebelled against everything Spanish. I wondered sometimes, seeing these mutilated images that redouble the Spanish preoccupation with suffering, about the limits of empathy across cultural lines. This *cristo* has drops of blood carved in the wood, running down its side.

In the summer of 1967 I was living in the Marikina Valley, at the edge of the Manila metropolitan area. A village had grown up along a lane branching off the main road, substantial concrete-block houses with tin roofs trailing off into huts built of wood and thatched with palm. Walking back and forth along the street to the household I had joined for the period of my research, I was gradually meeting my new neighbors, greeting and stopping to chat in one of the many tiny housefront stores.

Ordinary greetings in Pilipino, the national language of the Philippines, based on Tagalog, take the form of questions: "Where are you going?" "Where are you coming from?" As with the English "How are you?" however, which is not really a request for information, the polite answer, "Just over there," can be edged into richer conversation. Curious about me, children gathered to stare, and other adults drifted into the conversation. I was trying to understand how the community and the city beyond were seen by the villagers. Long-term residents near the highway, whose rural community was slowly being engulfed in metropolitan expansion, still spoke in terms of living in a *barrio*. Those who had recently migrated from the country to work in the shoe factories of the Marikina Valley maintained the mores of the province but saw themselves in the context of the city.

One day, in the late afternoon, when street life had

resumed after siesta, I stopped off to talk to a neighbor named Ana, and as we sat chatting we were joined by an older woman I had not seen before, Aling Binang, who had recently returned from the country. Ana had heard of the death of Aling Binang's twenty-year-old son some six months before and started to question her about it. After a while, Aling Binang began to weep, tears running one after another down her face, but the painful give-and-take of question and answer went on and on. I was careful not to interrupt the flow of a conversation both women seemed ready to prolong, but inside I felt outraged, very sorry for Aling Binang, and embarrassed by the tactlessness of Ana.

That evening, I found that I had to untangle two different reactions, writing a description of the conversation and, carefully separated, an expression of my feelings as a member of another culture. Each person is calibrated by experience, almost like a measuring instrument for difference, so discomfort is informative and offers a starting point for new understanding. Indeed, what I had seen and heard would not have pushed me to reflection and generalization were it not for the urgency produced by the sense of difference.

A few weeks later, a death occurred farther down the single street, where the houses were poorer and more rural in style and where my comings and goings had so far taken me less often. The family I was living with would be going to the *paglalamay,* "vigil" or "wake," and they urged me to come. I did my best to prepare for my part, asking a series of questions about what would be happening, and got instructions on how to give an *abuloy,* a "contribution" of less than a dollar, to the young woman whose mother had died. We went together to the house, staying until late in the night.

Again, I wrote two kinds of notes. One narrative describes the body laid out in its coffin, surrounded by *funeraria* lamps. The relatives had gathered, and neighbors were coming and going, expressing condolences and offering money and then standing and gossiping. Boys and girls were playing word games and flirting at the door, and gambling tables and barbecues were set up outside, with general merriment continuing through the warm night, noisily audible in the room where the body was laid out, overlapping and inter-

mingling. The other kind of notes concerned my own feelings: my reluctance even to go to this house, "intruding" on
the grief of others; my almost paralyzing embarrassment
over the act of giving the *abuloy*. Living in the village without
a husband and children, I was often grouped with younger
unmarried people; on that evening this meant joining them
in their noisy word games, laughing and imitating animal
noises, the equivalent of Old MacDonald had a farm, moo-
moo and cock-a-doodle-doo, only a few yards from a corpse
and a grieving family.

Because I had been briefed, I knew how to act, but my
feelings were foreign. For an American with a Protestant and
Anglo-Saxon background like mine, the handling of death
implies silence and decorum, requires that the privacy of the
bereaved be respected, and includes a certain reticence about
the material facts of every day. Later I understood that my
presence represented an extra honor for the old woman who
had died. The games and gambling were explained as necessary to keep people awake, to ensure that there would be no
solitude. In Pilipino, there is a word for "happy" which also
means "crowded," or "populous"; there is comfort in convivi-
ality. Wakes are important to young people, the best available
opportunity for courting.

Some months after I ended my research, back on the
other side of Manila, where Barkev and I were living, I
became pregnant. After weeks on my back fruitlessly hoping
to avert a premature delivery, I gave birth in a Manila hospital to the son we had planned to call Martin, who died a few
hours later. For me, the death of my baby was something
that should not have happened, unthinkable, unbearable.
But for the gentle Filipina nurses, the loss was sad but part
of life, bound to happen from time to time. Their sympathy
was firmly mixed with a cheerful certainty that I would be
back next year with another one—as so many women are in
the Philippines, whether the infant lives or dies.

It was our good fortune that my time in the village had
allowed me to observe and compare responses to death. On
the afternoon of Martin's birth, I described to Barkev the way
Filipinos would express their sympathy. Don't expect to be
left alone, I said, and don't expect people tactfully to avoid

the subject. Expect friends to seek us out and to show their concern by asking specific factual questions. Rather than a euphemistic handling of the event and a denial of the ordinary course of life, we should be ready for the opposite. An American colleague of my husband might shake hands, nod his head sadly, perhaps murmuring, "We were so sorry to hear," and beat a swift retreat; a Filipino friend would say, "It was so sad that your baby died. Did you see him? Who did he look like? Was he baptized? How much did he weigh? How long were you in labor?"

Stereotypes often conceal their opposites. In other contexts Filipinos describe Americans as "brutally frank," while Americans find Filipinos frustratingly indirect and evasive. Yet in the handling of death, Filipinos behave in a manner which Americans might characterize as "brutally frank" and seem to go out of their way to evoke the expression of emotion, while Americans can only be called euphemistic and indirect, going to great lengths to avoid emotional outbreaks.

Moments of deep loss or failure may feel as if they have no antecedent, as if nothing like this had ever happened before in human history. No one can know exactly how he or she will respond, what formulae will be seized, what words improvised. Yet individual responses follow cultural patterns, each experience offering analogies for others. Tradition even offered the materials for John Kennedy's funeral, although Jacqueline Kennedy had to combine them in new ways, familiarity both deepening meaning and offering comfort. But I was an outsider and the analogies I brought with me were off-key.

If I had not had the preparation of my time in the village, the most caring behavior on the part of Filipino friends, genuinely trying to express concern and affection, would have seemed like a violation. To avoid breaking down in the face of sudden reminders of grief, I might have imposed a rigid self-control, which would have reinforced the belief that many Filipinos hold, that Americans don't really grieve; or I might have reacted with anger to the affront, losing valued friends. An Iranian who had studied in London once described to me the revulsion he felt when his landlady mentioned to him unemotionally that her father had just died. So cold, from his point of view. So inhuman. But inhuman is exactly the wrong word. The potential for deep and profound difference is as distinctively human as the commonality that can be discovered beneath it.

Telling of a death, hearing of a death, expressing sympathy in the appropriate way, these are acts in which mutual recognitions of humanness are tested, but there is no single human way of responding. The bereaved is, among other things, a performer in a cultural drama that asserts basic ideas about the nature of life and death and the human heart. One of my students in Manila apologized for missing classes, because of the death of his grandfather. "I'm so sorry to hear that," I said. "Oh no," he said, answering the questions I forgot to ask, "we are all very happy for him; the priest was there and gave him the sacraments and now we know he is in heaven."

Some societies organize their recognitions of bereavement around an effort to help the bereaved regain control and forget, while other societies are geared to support the expression of grief. Some societies rehearse for grief and loss while others deny them. The most alien customs can be comforting once their rationale is understood, as an agnostic may be touched to receive a mass card, recognizing in the strange form gentleness, concern, the wish to help. Actually adopting unfamiliar customs in order to communicate is more difficult. I had been told how to offer the appropriate comfort at a village wake, but it was immensely—surprisingly—difficult not to feel a loss of authenticity in doing so. Curiously, one does not feel insincere translating words into another language, but translations of behavior come less easily.

Filipinos are fortunate in having a worldview which allows them to face the inescapable fact of death, including it in the rhythm of life and a continuing understanding of God's mercy. Americans treat grief almost like a disease, embarrassing and possibly infectious. For me in my own grief it was therapeutic to know that I did not have to fit American expectations, that it was all right to cry. Twenty-five years later, I can see that an awareness of these differences has stood me in good stead. What I learned in the

Philippines from Martin's birth and death prepared me for living and dying still to come; a village street, a wake, a hospital bed, encounters with grief in deepened understanding like a Fibonacci series.

Barkev and I had another protection from the pain and learning offered by unfamiliar forms, which was that after centuries of colonialism many urban Filipinos, like women and minorities, have become experienced at managing the contrasts that pervade their lives. They know that the behavior of foreigners may not be appropriate and regard the reactions of Westerners not so much as abhorrent but as erratic, so they try to avoid triggering unpredictable results. Those who overcome this reticence are likely to be those who most genuinely wish to be helpful. Friends came to each of us with their sympathy, and we were ready to recognize it and be comforted.

In recent years there has been a flood of discussion of the "other": that person or group that inhabits the imagination and, loved or hated, seems profoundly and significantly different. Whether negative or positive, the presence of the other leads to self-consciousness and puts familiar ways of being in question. Sometimes the other is the opposite sex, sometimes a minority group, sometimes even a distant culture described in terms that counterpoint one's own: Tahiti, darkest Africa, the mysterious Orient—all those regions whose strangeness is underlined to affirm the familiar. For a member of a dominant group, the sense of self is enhanced by a conviction of the inferiority of the other. Colonists may become more British or more French than they would have been at home. Any nation that has suffered or benefited from foreign occupation rapidly develops stereotypes and theories to explain behavior that seems bizarre. Such situations of sustained contact and contrast often find their own equilibrium and, in doing so, cease to be contexts for learning. Instead they become layered with rage and frustration.

A friend died when I was in Israel. This time I was given a briefing based on an awareness of both cultures and carefully warned about Orthodox burials so I would not be shocked at the thought of a body laid in the ground without a coffin. "We don't do it like the *goyim*," I was told.

Reminded that I was an outsider, I found myself drawn to the poetry of a genuine return to the earth. Often in Jewish history, the *goyim*, the gentiles, were illiterate and bigoted peasants, a reference point for differentiation. Sometimes, however, in periods of assimilation, being "like the *goyim*" beckoned, in spite of centuries of limiting influence and exchange. Israel, with immigrants from all parts of the Jewish Diaspora, is a melange of cultures, so a unifying contrast with non-Jews is important.

In Iran, Persian speakers agree in defining themselves as sharply different from Arabs, the foreign overlords of the far past, in spite of centuries of influence and borrowing in both directions. More recently, the prototypical foreign culture was different in different parts of the country. In Tehran, France once provided the model for emulation or rejection. In the south, near the oil fields, it was Britain. In the north, it was Russia. All these countries exerted influence, spreading in different ways across the country. *Qarbzadegi*, or "West strickenness," was a widespread disease. In Tehran there was a famous princely family in which the paterfamilias had had many wives and children, systematically sending his sons and daughters as they grew up for education in various countries and professions. Each became bicultural in a different way, and the family was positioned for survival in any change in the tides of foreign influence.

When we were in Iran, American models had become important, both as a beacon and as a threat. During the Islamic revolution ambivalence about the United States was crystallized in the imagery of the Great Satan. Yet when we arrived, Iranians in conversation with Americans were all too ready to disparage their own tradition, so I learned to press for the positive values behind statements like "We Iranians are too smart for our own good" or "too cynical ever to trust each other." Negative self-images are hard to shed without projecting hostility on the image of an inimical "other."

The Philippines has been under an even stronger American influence since the United States took over from Spain in the first years of this century. There was so much superficial Americanization that it was hard for new arrivals to get the early and vivid experience of difference which leads to learning, yet every descriptive statement about Philippine culture seemed to be implicitly contrastive. To write "Filipinos often take siestas," omitting to mention that they sleep at night as well, is to describe not Filipino sleeping habits but rather the differences between Filipino and American sleeping habits. It would almost have been possible to create the unstated assumption that Americans never sleep when it's dark by commenting, "Filipinos sleep at night." Worse still, such descriptive statements seemed to whisper the introduction "The trouble with the Filipinos is that unlike Americans . . . " Recent rising levels of self-confidence and assertion, as well as dawning deference from some Americans, may reverse the direction of the comparison: "The trouble with Americans is that unlike the Filipinos, they . . . " In other situations it may be "Unlike the British, Japanese, Native Americans . . . ," and so on. In a world of increasingly dense flows of information, "others" are all around, the hidden disparagements are revealed, the old forms of projection and distancing no longer work.

When I remember that first Feast of Sacrifice I observed in Iran, I am struck by the ambivalence of my hosts, differentiating themselves from their own traditions yet still drawn back into them. City people learn to imagine themselves in other shoes. Sometimes villagers see even their richer and more powerful countrymen from the city almost as another order of being, unaccountable and impossible to empathize with.

A certain amount of friction is inevitable whenever peoples with different customs and assumptions meet. It is familiar enough between genders or across class lines in a single society. What is miraculous is how often it is possible to work together to sustain joint performances in spite of disparate codes, evoking different belief systems to affirm that possibility. As migration and travel increase, we are going to have to become more self-conscious and articulate about differences, and to find acceptable ways of talking about the insights gained through such friction-producing situations, gathering up the harvest of learning along the way.

A few weeks before leaving Manila, I wrote an article for publication there, as an experiment in talking about the bicultural situation and the kinds of misperceptions it can produce. I used the sequence of experiences I had had with cultural attitudes toward death to illustrate the fact that complementary themes are found in every culture: true, Americans value frank speech, but in some contexts they are extraordinarily euphemistic; true, Filipinos deal gingerly with strong emotion, but in some contexts they evoke and intensify emotion.

The differences are not only in expression, as in using words or actions to convey sympathy, or deciding whether or not to allow oneself to weep in public. There are real differences in feeling as well. Because emotions are also learned, we are not the same under the skin. Filipinos and Americans, Iranians and Israelis have learned ways of looking and responding that can be very different. But there is a strange and special freedom that goes with entering another culture and entertaining different ways of feeling, realizing that they differ not in virtue or rationality but in cultural appropriateness. Comparing two cultures leads all too readily to regarding one as superior, so when I teach and write I try to compare at least three, sometimes drawing on my own experience and sometimes drawing on descriptions by others.

Along with mementos of Iran and the Philippines, I needed something for my MacDowell studio as a reminder of years spent in Israel, both the earliest and the most recent of my experiences of living abroad. I used to have another mandala, one of those diagrammatic plates marked for the ritual items of the Passover meal commemorating the exodus from Egypt: bitter herbs as a reminder of servitude, nuts and honey recalling the sweetness of liberation. Because it was eventually lost or broken, I represented it on my table with a paper plate marked in Hebrew into the appropriate sections, to represent a way of organizing and passing on the memories of emotion. Physical things are eloquent tokens of ideas, enriched by new meanings through time even when the tokens are no more than evanescent paper representations. Often material objects turn out to be diagrams, cognitive maps, that share our space, teach our children, and argue for ways of organizing experience. Like a shell that encodes its own process of growth, objects summarize histories. Passed from hand to hand they represent new relationships and meanings on each passing.

My senior year of high school in 1956–57, which I spent in Jerusalem, was the first time I lived for such a long period in another country. The senior year of high school is a difficult time for encountering another culture, for adolescents tend to be preoccupied by the need to belong and to reject those who are different. Nevertheless, Israel offered me at sixteen an intoxicating opportunity to immerse myself in another language and culture and to experiment with models of belonging and community. After the Sinai campaign in 1956, when Israel had occupied the Sinai desert for the first time, a group of my classmates set out with a large number of others belonging to the same youth movement for an extended hike over the Hanukka vacation, into the wilderness where the liberated Jews wandered for forty years after leaving Egypt. Friends suggested that I come along.

The area was bleak and mountainous, rock desert rather than sand dunes. Because of the Biblical background, Israelis treat such a trip as a voyage of self-discovery, not so much an encounter with wilderness as an encounter with history. We went south by truck to the staging area where we were to set out on foot, each person carrying a backpack with food and water for three days and a sleeping bag. Gathered at the foot of a steep hill, with a narrow trail winding up between the boulders, we were warned that once we left the trucks there would be no turning back and that no one who could not last the course should go.

Worried, I started up the hill, but I had had nothing like the hiking experience of my friends, and this was the first time I had tried to walk or climb with any significant weight, so halfway up that first hill, with the trucks still waiting at the bottom, I told the other girls from my school that I could not be sure I would be able to keep going and it would be unfair to the others if I stayed in the group. I stepped off the trail and sat to one side, prepared to start down when everyone had passed. Along came a group of the boys, insisting that they could take turns carrying my pack and I would surely be able to keep up with nothing to carry, and I let myself be persuaded by their enthusiasm.

Within a couple of hours, too late to turn back, it became clear that even without a pack I was in trouble. Regular rest stops were called, and each time we stopped I lay down and fell asleep for the few minutes available. There were magnificent views, but I did not see them, staring at my feet and stumbling from fatigue. The day ended with the celebration of the first evening of Hanukkah, with lights lit in tin cans and the group singing folk songs. I managed to keep going until the afternoon of the second day, to a point when I couldn't get up. By then we were on a well-marked trail to that night's resting place, so two of the boys stayed behind with me for an extra hour and then we walked the remaining distance, I leaning on their shoulders, arriving after dark.

The next day I started out walking on my own again, and the going was not so steep. At midmorning, someone asked me whether I would carry his camera and step out of the line from time to time to take pictures, and I felt absurdly honored to be able to contribute in some way to the group. Euphoric with fatigue, I tried to sort out my feelings, with two quite different sets of emotions running simultaneously. The Israeli sequence was informed by the socialist ideology of the youth movement, with its emphasis on solidarity, mutual help, and commitment to the group. As long as I tried my hardest it was not inappropriate to receive help, but Israelis sometimes give very short shrift to those who do not do their best. At the same time, I could review my emotions as an American teenager, feeling humiliation and resentment of those who had helped me and discovered my weakness. Part of me felt deep love for and closeness with my comrades. Part of me wanted to withdraw and avoid them in the future.

Beyond my feelings about my collapse, I found myself thinking in two ways about the question of whether it could be right to risk burdening others and about the meaning of helping. Training patterns in the Israeli youth movements and in the army are said deliberately to create situations in which members of the group learn to take responsibility for one another and to share burdens. Just as my experiences in the Philippines gave me a second way of thinking and feeling about death, my experience on that hike into Sinai has informed my feelings about all those circumstances in which one part of a community supports another, providing education or welfare or health care. Receiving help can be bitter, a shaming reminder of inadequacy, but that experience taught me that real help does not treat need as the result of irresponsibility or malingering and is generous enough to make it possible to contribute in turn.

I have often chosen to go into unfamiliar settings in spite of the discomfort involved, gaining a sense of perspective in my life that has a very different kind of value from the production of books and articles. Still, I have wondered how I would have reacted to Martin's death in Manila or my failure to keep up with my Israeli friends if they had occurred soon after arrival in a strange environment. It is not easy to use the crises of one's own life as the stimuli for new ethnographic insights, yet we all arrive as strangers at the moments of crisis in our lives, having to improvise responses from previous learning. This must be labor; this is bereavement.

Arriving in a new place, you start from an acknowledgment of strangeness, a disciplined use of discomfort and surprise. Later, as observations accumulate, the awareness of contrast dwindles and must be replaced with a growing understanding of how observations fit together within a system unique to the other culture. Having made as much use as possible of the sense that everything is totally alien, you begin to experience, through increasing familiarity, the way in which everything makes sense within a new logic. Eventually an ethnographer will hope to develop a description of a whole way of life that will convey this internal consistency, in which the height and placement of a chair, the adult response to a crying baby and to voices raised in dispute, and the rules about when to relax and the rhythms of the day can be integrated, although never perfectly. The final description should deal with the other culture in its own terms. Yet it is contrast that makes learning possible.

When I first arrived at the MacDowell Colony, I put up a picture in my studio of a row of big brown bats on a calendar Vanni gave me and a clipping from the Sunday *Times* about crashing amphibian populations, sent by my editor, who recognized another of my concerns, and I set out other mementos around the room. The old verse about what a bride should wear to her wedding is a good rule for all transitions, "Something old and something new, something borrowed and something blue." At one time, I thought the blue was there only to provide a rhyme. Then I wondered whether it was included as a reminder of the sorrows and failures that lurk within any new commitment. Once the word was set into the tradition, a dozen interpretations would have been invented, in the human habit of seeking for meaning. Today on my worktable the blue and green of the miniature globe stand for the integration necessary for life, waters where the ammonites propelled themselves long ago, the ponds drying up and leaving frogs and salamanders bereft. The rhyme suggests not miscellany but the complex spiral of exchange from generation to generation, replication and recombination.

3

Double Helix

THERE WAS A LOT OF SNOW in New Hampshire during the winter I was at the MacDowell Colony, turning the route to my studio into Heraclitus's river, different at every passing. As my writing moved along, forage became scarce in the woods, and I often saw deer. Each return over the same ground represented layers of change: in me, in my manuscript, in the landscape.

Even what appears to be a repetition is often a return at the next level of a spiral or, more mysteriously, the other side of a Möbius strip. Take a narrow strip of paper about six inches long and give it a single twist before taping the ends together. Then start anywhere to draw a continuous line along the surface. The first return will be to the opposite side of the paper; only on the second full round will you meet the beginning of your line, "And know the place for the first time." Mathematicians call this a one-sided surface since the entire surface can be covered without ever lifting the pencil.

Barkev reminded me of the Möbius strip when I started talking about spirals, so I made one to put on the table with my other visual aids, a new-old desk friend constructed from paper. My desk was becoming crowded, not by these few objects but by the range of meanings they evoked. Much of my writing consists of taking ideas that are coiled within one another. Before spinning and weaving, wool must be carded, and in the same way thoughts must be opened into sequential prose. It would not do to lay them out too precisely, however, for I have wanted to convey something of the process of learning, and most learning is not linear. Planning for the classroom, we sometimes present learning in linear sequences, which may be part of what makes classroom learning onerous: this concept must precede that, must be fully grasped before the next is presented.

Learning outside the classroom is not like that. Lessons too complex to grasp in a single occurrence spiral past again and again, small examples gradually revealing greater and greater implications. The little boy staring wide-eyed at the sacrifice of a sheep may one day be a *hajji,* one who has completed the Meccan pilgrimage and seen the sacrifices and the Holy Cities and returned home looking at ordinary life differently. The effect of such partial repetition is to heighten contrasts, sharpen the differences created by context. A son will experience the Feast of Sacrifice differently as his father ages, discovering new pleasure if the family becomes prosperous enough to buy its own sheep. Through the years family constellations will shift and the society regroup as his own body learns about strength and illness, sex and dying. Morals are rarely drawn, and the comparisons are not made explicit, but anyone who has wept at a wedding knows that the past and the future are present in each single ceremonial.

In the same way, the stories told here all have more meaning than I know how to unpack in the context of a single chapter, and the objects on my desk will shift to represent different concepts. It may be a good idea to begin reading from the beginning, but the reader who returns from the end to the beginning will find "a ring when it's rolling" that has no end. We will not finish with the Persian garden in one visit, or the Passover meal, or the Filipino rage at arrogant Spanish noses. In recent years, I have been learning not only about improvisation as a mode of participation and observation in the present but about the possibility of recycling the past, the flashes of insight that come from going over old memories, especially of events that were ambiguous, mysterious, incomplete. In the past, when memorization was a common form of learning, children committed long passages of poetry and scripture to memory without understanding them. Then, if the texts were well chosen, they had a lifetime in which to spiral back, exploring new layers of meaning. What was once barely intelligible may be deeply meaningful a second time. And a third.

Spiral learning moves through complexity with partial understanding, allowing for later returns. For some people, what is ambiguous and not immediately applicable is discarded, while for others, much that is unclear is vaguely retained, taken in with peripheral vision for possible later clarification, hard to correct unless it is made explicit. Beyond the denotations lie unexplored connotations and analogies.

What we call the familiar is built up in layers to a structure known so deeply that it is taken for granted and virtually impossible to observe without the help of contrast. Encountering familiar issues in a strange setting is like returning on a second circuit of a Möbius strip and coming to the experience from the opposite side. Seen from a contrasting point of view or seen suddenly through the eyes of an outsider, one's own familiar patterns can become accessible to choice and criticism. With yet another return, what seemed radically different is revealed as part of a common space.

If I want my students to be able to observe their own culture, I offer them alternative versions of the same sequence, sufficiently unfamiliar to focus their attention. One of the best ways of doing this is by looking at familiar patterns of growth framed in other cultures, so I may ask them to read life histories like the life of the Winnebago Indian Crashing Thunder, or the life of the San woman Nisa, or to interview acquaintances with different backgrounds. In Iran, I carried this approach a step further by bringing, on separate occasions, American and Iranian mother-infant pairs into the classroom. Part of my purpose was to show that very small infants learn some of their assumptions about the world even before learning to speak. Many of those assumptions are then never put into words but are retained at such a deep level of learning that they seem self-evident. My other purpose was to persuade my students of the legitimacy of learning from observation.

I brought an American pair first, so the students' eyes would be opened by strangeness, and then an Iranian pair. The mothers were both well-educated, affluent homemakers with first children, daughters of about ten months, able to get around the room but not yet walking or speaking. I had told each mother that the students would be observing her infant and had asked her to bring whatever she would need to keep her daughter content for the two-hour class period. I set a chair and a light cotton rug (*zilu*) on the floor at the front of the room.

The American mother, we'll call her Joan, sat on the chair and set her baby, Becky, on the floor beside her, on the rug. Becky was immediately fascinated by the watching students, as they were fascinated by her bright-eyed responsiveness. She wriggled on her belly to one after another, responding and crowing as they dangled pencils and keys and offered her candy, climbing up their legs and flirting. From time to time she looked back at her mother, who nodded and encouraged her explorations. The floor was filthy, and Becky's hands were quickly blackened, her jumpsuit and face smeared. After a while, her mother fetched her, scrubbed a bit at her face and hands with a paper towel she moistened from a thermos, and offered her some toys—a set of colored cups that fit one into the other, a puzzle, and a hard-paged book. "Apple," she said, pointing it out to Becky. "See, the apple is red. Button," on the next page. "A blue button." Becky played for a while with the toys, then ventured out again to make friends with the students.

When the class came to discuss their session with Becky and Joan, they were unanimous in their delight with Becky and their criticism of Joan. She had failed, they said, in her responsibility of keeping Becky clean; Becky could have picked up all sorts of germs from the classroom floor. One of the topics that had come up repeatedly when I interviewed foreign women married into Iranian families was the freedom that Iranians feel to approach a woman in public and criticize the way she is treating her child. "That child isn't

warmly enough dressed," they used to say to me about
Vanni. "That child will get dirty." The business of child rearing is not private. Strangers accompany their advice with
food offered without checking with a parent, or kisses on the
lips. Vanni loathed having her cheeks pinched, but Iranians
were charmed by her responsive openness, and disapproving
of our permissiveness. In a supermarket, she captured the
manager and ended up on his lap eating ice cream, which
became her regular prerogative. In a restaurant—there are
often children in restaurants and nightclubs in Tehran, sitting quietly in their mothers' laps without high chairs and
gazing wide-eyed at the crowd until they fall asleep—Vanni
went up to the stage and joined the chanteuse. Unaccus-
tomed to more gregarious American children, Iranians are
not armored against them.

A week after Joan and Becky's visit, I brought to my class
an Iranian mother and child, Parvaneh and baby Shahnaz,
almost the same age as Becky but just slightly ahead in
development since Shahnaz had learned to crawl rather than
wriggling from place to place on her tummy. Parvaneh
ignored the chair and sat down on the floor with Shahnaz,
modestly tucking her full skirt around her legs. She was not
veiled but wore a silk kerchief. She gave Shahnaz her own
handbag, with handles that flipped back and forth, to play
with, and a rattle and a small doll she had brought with her.

The students had carried over expectations from the previous visit and were looking for the same response from
Shahnaz that they had gotten from Becky. They cooed and
called and offered whatever enticements they had in their
pockets—keys, a candy, a string of worry beads—but each
time Shahnaz crawled to the edge of the rug she stopped.
Her mother reached out with an enclosing gesture, patting
her and telling her, go for a walk, go see the nice students,
but offering her a piece of banana at the same time.

After a while, one of the students, frustrated, came and
took the doll, standing it up just a few feet beyond the edge
of the rug, with the mother's purse strap over its shoulder
and his own shiny key ring in its hand. Shahnaz crawled to
the edge and gazed. You could hear her breathing quicken at
the fascinating assemblage of toys. She literally panted with

desire. And then she *turned her back* on the room to play with her mother's foot. The students later agreed that Parvaneh was a very good mother, pointing out that in the course of the class session she had produced a bottle for Shahnaz while Joan had neglected to bring food, and that Shahnaz's lacy dress had remained clean and fresh.

Of course I was lucky to happen on a pair that offered such a sharp contrast. But I did the same thing a year later with different mother-infant pairs and the contrast was only slightly less vivid, and I had been watching the same patterns in families we visited and in the way people responded to Vanni. Infants go through stages in their response to strangers, but parental expectations continue to shape them. Each culture has different expectations of male and female infants, but a comparison of mother-son pairs would have exemplified the same kinds of differences in the handling of space and exploration.

While the students were watching the mothers and infants, I was watching the students and the entire classroom scene. The door of the room had a peephole in it, and I was constantly aware of movement in the hall, one eye after another peering through the window. Something unheard of was happening in that classroom. Classrooms in Iran are places where professors lecture to students, who are expected to memorize what they are told, combining French models of pedagogy with the traditions of Islamic scholasticism. Iranian students are highly sensitive to interactions and skilled at observing their professors, working out what will please them, and negotiating their grades, but these skills are not seen as internal to the educational process. Observation—particularly observation of an infant—simply did not fit the definition of what is appropriate to educational settings. Education is indissolubly linked to authority. Indeed, there was one man in the class, a military officer always in uniform, who gazed out the window the entire time. I wondered sometimes whether he was the SAVAK spy I was told to expect in every class. In any case, what was going on with mothers and infants was beneath his attention.

If you try to untangle the actual patterns of learning and

attention in that classroom, some interesting paradoxes emerge. At ten months of age, before speaking her first word, an infant knows a great deal about her world and what is expected of her. In some sense, Becky knew she was supposed to demonstrate independence. Shahnaz knew she was supposed to stay close to her mother and avoid strangers. She recognized that the flimsy cotton rug on the floor established a geography in the room, a boundary to the contained and hospitable space of the familiar that it was inappropriate to cross. Shahnaz could recognize the very minimal evocation of these rules in a thin cotton rug, which Becky simply ignored.

The mothers, unself-conscious about their own behavior, wanted their daughters to show up at their best. Parvaneh was giving Shahnaz contradictory messages, one with her restraining hand and the offering of food, the other with her words, urging Shahnaz to go join the students. Iranian courtesy is full of invitations that are not meant to be accepted, but Iranians enjoy the tension this creates, and Shahnaz, whether or not she yet understood her mother's words, already understood *how* to understand them. Joan explained to me that she had selected typical American toys, which were in fact all "educational." They were also all beyond Becky's developmental level. She could enjoy handling the pieces of the puzzle but not fit them together, and it would be months before she could learn that an apple is "red."

Each mother was underlining fundamental lessons by her behavior, but these were not necessarily the lessons she would have said she was teaching: the Iranian mother was urging sociability even as she was emphasizing a bounded world of warmth and nurturance, surrounded by danger; the American mother was using educational toys to project exploration and competition. The mothers, of course, were improvising, for I had put them in a novel and somewhat embarrassing position, which they cheerfully accepted, using habit and common sense (both learned) to shape their behavior.

The boundaries that little Shahnaz had learned to accept reminded me of a mental topography that gradually became clear in the Philippines. Like many languages, Pilipino has

three kinds of demonstrative: this (here), that (nearby), and
that (distant); here, there, and way-over-there. Wherever we
went, we found we were warned of danger outside or beyond
some perimeter. Around Manila, on the island of Luzon, we
were told of the risks of travel to the big southern island,
Mindanao, but when I told people I met in Mindanao of a
plan to travel to the Sulu Archipelago, the scatter of small
islands that trails off the southern end of the Philippines,
they warned me to take care. For each island and in every
Manila neighborhood I spent time in, there was an area "out
there," threatening and unsafe, providing a contrast to the
safety of the familiar.

Some Americans living in the Philippines simply synthesized these warnings to reach the conclusion that the entire
country was unsafe, failing to notice that the warnings given
by Filipinos, including lurid stories of crime and violence,
implied a contrast between an area of safety and an area of
vulnerability. Foreigners often made the same kind of mistake in Iran, listening to the cynical remarks of Iranians
about others—particularly those in the public eye—and synthesizing these into the conviction that an entire people
could be always devious and insincere. We fail to hear the
implied comparisons: dangerous compared with what? insincere compared with whom? Outside danger may be a part of
the comfort of home.

The two infants, still limited in their interactions by the
lack of language and only recently mobile, traced alternative
theories of social life onto the floor of that dusty classroom,
a topography of value and safety in space and human inter-
course. The mothers too had quite different theories about
what infants can and should learn. In Iran, children are characterized as innocent. Even in this society of sexual segregation, mothers traditionally took their sons with them to the
bathhouse, where they would be surrounded by scantily clad
women and sexual gossip, until they were visibly approaching puberty. Adult men remember listening and watching
avidly, protected by the assumption that they did not understand. In the same way, while his mother is drinking tea and
chatting with a guest, a three-year-old dives for the bonbon-
nière of nuts and dried fruits, knocking it to the floor and

breaking it. "He's only a child, he doesn't understand" is the adult comment.

Westerners have tended to attribute purposes to children: "He's trying to get attention," "He's always into mischief," or even "'He only does it to annoy, because he knows it teases.'" Western child rearing suffered for a long time from the idea of original sin, in which badness is inherent and must be overcome by discipline and training. Bad behavior in a child used to be attributed to native waywardness permitted by parental laxity. More recently, misbehavior has been blamed on mistreatment or misteaching, with exploration seen as a positive value. Iranian children are good but not knowing. For our Puritan ancestors, children were knowing but not good.

The watching students, meanwhile, were being confronted with a quite radical challenge to their habits of learning and attention. They reacted to the mother-infant pairs according to their own assumptions and common sense, at first judging the behavior of mothers and infants separately; but my purpose was to challenge them to think about the connection between the behavior of the mothers and that of the infants, the unstated assumptions and values the mothers were teaching, invisible bonds stretched in the air. It was only when they were offered a contrast, a moment of strangeness followed by the discomfort of having me point out their contradictory pattern of approval and disapproval, that they could begin to see that there was something to be discussed beyond a simple matter of nature.

I think it would have taken a series of examples, such as they would have seen by living in another country, before they would have fully taken in what I was trying to show them, all at the same time: a new concept of classroom learning simultaneous with a new understanding of the working out of patterns of nature and nurture in the interactions of mother and infant. Still, they will remember and retain at least one layer to build on, one that would not show up on an "achievement" test. Without fully understanding what they were seeing, they were also drawn into a shared performance, as we all were in that dusty classroom, for one another and for the watchers in the corridor. With luck, life will provide a return of the spiral, to be met with recognition rather than bewilderment.

Later, when I was teaching at Damavand College, an American-sponsored college for women where I had some married students, one woman asked me what I did when my child had a temper tantrum. "What do you do with yours?" I said. "Well, I try not to give in, but then when it gets really bad, I just give him what he wants." "Exactly," I said. "Why not give him something else to think about before he gets all worked up?"

My answer made no sense to her until the following week. We had begun holding classes in a new set of buildings, beautiful but uncomfortable and difficult to reach. The administration improvised a shuttle but only announced a fare after it had been running free for some time. The students planned to gather on that very public corner, refusing to pay to board the bus and working up to a shouting demonstration that might attract the police—a temper tantrum that administrators would quickly give in to. But at seven o'clock in the morning, on the day of the new fares, with snow on the ground, I was the first one out on the corner, with a pocketful of change, bantering and joking with the students and going through elaborate Iranian courtesy routines. When I commiserated about the fares before they could protest and paid for several of them, they were deflected from protest to protestation. After three or four days of "After you, Alphonse," fares on the shuttle were taken for granted. My student came up to me and said, "I was watching you. Now I know why your baby doesn't have temper tantrums."

As an American parent, I had learned to defuse certain situations by deflecting attention, but the real advantage that I had on that street corner was that of knowing two sets of rules and selectively combining an Iranian behavior with an American one in a way that would not have occurred in either culture. Although there are ethical problems in using cultural knowledge in this way, resolving the conflict seemed to me to justify it.

What Becky and Shahnaz had already learned about what to value, where they could safely go, and who to inter-

act with had established a framework for their future explorations and so for future learning. One day at a student gathering at our house, a woman in one of Barkev's classes
invited Vanni, age three, to stay overnight at her house.
Vanni went and got her pajamas, and I found the two of
them standing, embarrassed and awkward, by the front door.
It had never occurred to the student that a three-year-old
would be willing to set off to spend the night with a near
stranger. It had never occurred to Vanni that this friendly
woman, a guest of her parents, would extend an invitation
that was not intended to be accepted. Should I tell Vanni that
the invitation was false? Insincere?

"Sweetie, Mahin invited you to her house as a way of saying how much she likes you. But she didn't expect you to
actually come. She probably isn't set up for overnight guests.
Besides, you both have to go to school tomorrow." That
much had to be said; my problem was to find the tone to
convey the fact that Mahin was entirely sincere in the
warmth of her invitation but not in its substance. The true
message depended on a combination of her words, her
understanding of the context, and what she took for granted
about human nature. The explanation would have been
unnecessary with an Iranian child. I wanted Vanni to retain
her freedom but to put it in cultural perspective.

If you ask Iranians about Iranian national character, they
will often speak of distrust or cynicism, two words that
translate the Persian *badbini*. This is a traditional negative
orientation to unfamiliar places or persons or ideas. I used to
warn new arrivals in Tehran that they might be shortchanged
in little neighborhood shops: viewed as predatory, they
would be objects of predation. But if they kept returning,
they could gradually become part of the community. Within
the circle of trust, to those defined as insiders, Iranians are
extraordinarily trusting, but this is obscured when they are
describing their culture to Western researchers—outsiders.
Like the stereotypes I heard in the Philippines, this stereotype conveys only one side of the coin. And, as in the Philippines, the negative stereotype generated in this way may be
played back as self-condemnation. It is Iranians who will tell
the foreigner never to trust anyone.

In small-town America today, most people still leave their doors unlocked. For all the folklore about disruptive outsiders, new arrivals are not automatically categorized as dangerous. But levels of diversity and miscommunication are rising all over America, gradually shifting people into the defensive modes you can see in the cities, where strangers are guilty until proven innocent. Even as I write this, some twenty years away from having my own child to protect, I reflect that the sense that children are under threat has become so strong in America today that I would probably inhibit a child's exploration in new ways, ways that would surely convey a habit of suspicion, not only as a situational adaptation but as an element of character. Learning at that profound level shapes all subsequent learning—a one-year-old's attitude toward strange adults, for instance, is recycled in attitudes toward teachers, acting as a filter for the opportunities of school.

Experiences spiral through the life cycle, presenting the same lessons from new angles: parenthood offers a new view of childhood, so does grandparenthood and so also the roles we are sometimes offered in relation to the children of friends. Observing infants, strange as visitors from another planet—and the even less intelligible infants in other cultures—is one way of making one's own early learning accessible to awareness and change. Other lessons of childhood are examined and reappraised in psychotherapy. It is possible even for childhood to be twice-learned, seen from the outside as well as from the inside.

I arrived in Iran from a period of research on mother-infant communication. When Vanni was born in 1969, I had decided that since my attention was going to be divided for some time to come, I would arrange my professional work so that the resonances of research and motherhood would be a source of insight, so I started working at the Research Lab of Electronics at MIT, where another researcher, Margaret Bullowa, had collected films and tapes of mother-infant interaction. I used to sit viewing those films for hours, dashing home to nurse Vanni when the baby on the screen reminded me that it was time, hoping to document from the films behaviors I was responding to with my daughter. And indeed

I found myself describing a pattern of playful vocalizations between mother and infant that I spoke of as the epigenesis of conversation. My doubled experience of observing my own infant and the infant in the films was a first step in understanding that participation precedes learning.

What interested me about these "conversations" was that whereas the mother was speaking sentences to the infant, using the words and grammatical patterns of adult language, the infant was responding with little coos and gurgles. The internal structure of the infant's behavior was quite different from the internal structure of the mother's behavior, but together they were collaborating in sustaining a joint performance. That capacity, and the delight it gave them both, preceded the differentiation of linguistic structure—learning words or grammar or rules of sentences. In the same way, in the later scene in the Persian garden, Vanni was able to be a participant in a ritual with a complex history and meaning which she knew nothing about. The babies who visited my classroom, not exactly university students, were able to play roles already shaped by culture; classrooms too involve joint productions with only partially overlapping systems of meaning.

It is hard to think of learning more fundamental to the shape of society than learning whether to trust or distrust others. Little Becky will probably move as an adult some distance from the nuclear family she was growing up in, and will repeatedly need skills in establishing new relationships. Little Shahnaz was learning to withhold trust except from those who were closest—family, siblings, long-term friends—but she came from a considerable network of kin. The ideal marriage in traditional Iranian culture is with a cousin or, if not a cousin, a member of the same community.

One of the things I treasured in Israel was the determined assertion that community existed even before efforts to build it, although Jews from different countries had deeply different cultures as well as the sprinkling of criminality and pathology to be found in any society. I remember in 1956 visiting a large and long-established kibbutz, with a factory in addition to farming, and being startled to be handed a key to my room, because "there are so many people one

doesn't know"—startled only because until then everything
had seemed so open. The best airport security system in the
world still relies on the assumption that Jews do not blow up
airplanes. As a sixteen-year-old, I lived alone in Jerusalem
and went for long walks, far into the night. I hitchhiked all
over the country with friends, striking up conversations in
the street. Community, like the sacred, is an idea that
becomes reality because we believe in it, not vice versa. In
Israel, many families skipped a generation in which children
grew up without grandparents present, but today, with the
aging of a new generation, the grandparents are there again,
for the idea of treasured and respected grandparents has survived.

The lessons of school gain authority because they are layered onto earlier informal learning in the home, which is
where we learn how and what to learn and how to transfer
knowledge from one situation to another. These vital skills
mostly remain outside of awareness. Looking at how much
has been learned within a few weeks of birth proposes a new
kind of respect both for nonschool learning and for the
capacities of the very young. Children and traditional peoples, even illiterate peasants who have had enough exposure
to the city to regard themselves as profoundly ignorant, have
vast amounts of knowledge long before teachers and social
reformers get to them. An awareness of the complexity of the
knowledge they already possess could in itself be a revolu-
tionary force.

Not only do we not know what we know, we don't know
what we teach. All societies pass on complex patterns: conventions of human relations; languages roughly comparable
in their basic complexity, whether or not they have ever been
written down; details of the environment; skills for survival;
abstract notions of causality and fate, right and wrong. Some
are learned by play and experience, as when a little boy takes
his own dugout canoe out on the reef; others by observation,
watching the grown-ups operate the elevator in an apartment
building. Everywhere there is some teaching, mostly by family members who, like Parvaneh and Joan, are not teachers
and have no systematized knowledge of the material they are
passing on. There is tremendous variation in formality, in the

demand that a performance be correct the first time, in styles of reward and punishment—"He's just a child; he doesn't understand"—and the list of individuals who are allowed to play an active role.

My classroom at the University of Tehran, where the students were confronted with strangeness and asked to behave in unfamiliar ways, is a metaphor of the world we live in today, pressed by change and by contact with other ways of thought to question premises learned even before language. The students knew, almost as deeply as Shahnaz knew the meaning of the transition from carpet to floor, without ever having put it into words, what was supposed to happen in classrooms, and that knowledge limited their capacity to learn by observation.

The problem of an outsider as teacher is to enrich students with new learning skills, not to replace the old ones, and this demands an awareness of differences. At Damavand College, American teachers struggled to convey their rules against plagiarism without being aware that the notion of ideas as property that underlay them was itself foreign and unfamiliar. They believed they had no wish to disrupt the society they were working in, yet they tried to persuade students that it was appropriate in the classroom to question the expertise of the professor. On the other side of Tehran, my husband and his colleagues were teaching students of management to learn from case discussions rather than by memorization. The most basic assumptions are rarely made explicit by either teachers or students.

If children learn, even before words, that the unfamiliar is inimical, this will affect their approach to differences of all kinds, even those forays into the unfamiliar that we take when walking into a forest or a meadow, and they will never be comfortable in unfamiliar social worlds. If they learn that their way—or any single way—is always best, they will never see, and use, the alternatives, however widely they travel. To get outside of the imprisoning framework of assumptions learned within a single tradition, habits of attention and interpretation need to be stretched and pulled and folded back upon themselves, life lived along a Möbius strip. These are lessons too complex for a single encounter, achieved by garnering doubled and often contradictory visions rather than by replacing one set of ideas with another. When the strange becomes familiar, what was once obvious may become obscure. The goal is to build a complex structure in which both sets of ideas are intelligible, a double helix of tradition and personal growth.

I was a schoolgirl when I first went to Israel, a young wife when I went to the Philippines, and a new mother when I went to Iran. The landscape and I were both different at every turn of the spiral, almost decades apart; because it is impossible to step into the same river twice, one can learn from each return.

4

Something Blue

RECALLING CHILDHOOD makes it possible to experience it again, to discover another way of seeing within one's own skin. Even though children begin very early to fit in with adult expectations, they continue to be enigmatic strangers, so the visions of childhood could be treasured as alternative ways of seeing. The perceptions and experiences of childhood continue to be visible out of the corner of the eye in daydreaming, in free association, and in sleep. Some become the defining foundations of later learning, built up into the shared understandings of society, while others are disallowed in the adult culture, so that not only episodic memories but whole modes of consciousness are buried.

When I wrote a memoir of my parents, *With a Daughter's Eye*, I found myself able not only to recover a wealth of memories but also to reconstrue those memories with the help of adult knowledge, to infer the motivations underlying mysterious adult behavior, to empathize with the adults in their puzzling world and with my earlier self. There is today a whole range of psychotherapies based on the idea of communion with an inner child, perhaps in need of comfort or perhaps, like Wordsworth's, offering intimations of immortality. We could equally speak of communion with parents long gone, through maturing identification.

I was encouraged as a child to write stories and poetry,

which I dictated to my mother. Once upon a time, I said in one of these stories, there was a sad, dreary kingdom that had no colors. Everything was black or white or gray, gray skies, a white-hot sun, black leaves. Even the flowers, although they had beautifully shaped petals and fragrant sweetness, seemed to the people of that kingdom to be no more than different shades of gray. Although the people had no idea that it could be different, still they lived gray and joyless lives. That was just the way things were.

But after a while the king and queen of that kingdom had a baby girl, and she seemed different from all the other children of the kingdom, who were solemn and docile. She was playful and full of laughter and curiosity. By the time she was a few years old, everyone knew there was something special about her, and she herself began to understand that the people around her, whom she loved, did not see the world with the same eyes of happiness. Gradually, too, she learned that in many cases where she saw differences other people saw sameness. Two kinds of flowers that they saw as the same she saw as different. The same was true of birds and butter and books and bedspreads. In fact, she was the only person in the whole kingdom with color vision.

At first she felt that she must be in the wrong and all the others right. She was a good child and had grasped early on the responsibility her parents and teachers felt to explain the world and bring her up to be a queen, so she tried to behave herself more sedately, but still she sometimes couldn't resist crying out with delight. She dressed with a shudder in the clothes laid out for her by her nurse, and tried to be polite about it. It took many years before it occurred to her that she might know more than others, and then that she might be able to make her vision of the world available to them. What would it be like if all these dour people also saw the world as brilliant and sparkling? The grown-ups, busy about her education, never even thought of learning from her, although of course the servants had to listen politely since she was a princess. Even younger children shook their heads in bewilderment when she talked about what she saw, and often she floundered for lack of words.

One day she stopped asking who might be able to imagine that she knew better and simply asked herself who loved her best. She went and sat in the queen's lap and said, "Look into my eyes and tell me what you see." "Why, you have beautiful gray eyes," said the queen, "as pale gray as egg yolks or lettuce leaves." "No," she said, "look more deeply." They sat for long minutes, and finally the queen said, "I see something I have never seen before. And what I see is different from egg yolks or lettuce leaves." So the princess said, "What you see is called blue. If you look now at an egg yolk, you will see that it is yellow, and the lettuce leaves are green." (The six-year-old making up this story is female, blue-eyed, and striding back and forth dictating authoritatively to her mother at the typewriter, but the words of this version are those of an adult who remembers only the plot, not the language.)

The princess took her mother by the hand and, leading her around the palace and the palace gardens, taught her to see color. Then they went to the king and taught him to see color, and bit by bit everyone in that kingdom learned to see color. Indoors and out, they all started laughing aloud at how much more variety they could see in the world than they had ever imagined.

The premise of the story breaks down at the end, for as a child I never quite penetrated the princess's dilemma. Somehow the word *blue* and the concept of color are ready to hand as soon as the loving magic transforms the mother's vision, but children have no words for their private knowledge. Perhaps I should have imagined the process of inventing names for the new experience, words for colors that were really names of objects, like *orange,* or a scholarly debate about the nature of the new experience, something like the transition that took place, within a few years of the vivid experience of Jesus' friends, to the invention of trinitarian theology. I seem to have understood that for the individual color is an artifact of perception, but it didn't occur to me as a child that color is encountered through the understandings of a community. Whatever the physics of wavelengths of light and the biochemistry and neurology of color vision as determined for most of us by genetics, the experience of blue remains totally private, and the shared meaning of blue has to be socially constructed—not, of course, from nothing but as a building is constructed from the facts of gravity and balance, the characteristics of materials, the purposes and meanings of human beings.

The fable expresses a reversal of the most visible patterns of education. Here, a child teaches adults. That reversal occurs more often than we realize, for much of what adults learn is learned from children. Newborns come equipped to turn their elders into parents, and on-the-job training continues from then on. After immigration or when the environment is changing rapidly, parents must use their children as guides to the new reality that surrounds them, like the wise children in fairy tales, or as models, as the Gospels demand. Learning is ideally a two-way street. Children participating in adult occasions see things adults have learned not to see and guess at meanings missing in official explanations.

Several years ago I was invited to a weekend conference at the Cathedral of St. John the Divine in New York to explore emerging understandings of human perception and to wrestle with the new biology in relation to Buddhist and Judeo-Christian understandings of the interface of cognition and reality. I had been asked, as a layperson, to preach on Sunday to the normal Sunday congregation with conference participants mixed in, so I would have to speak at two levels. At one level, I would be responding to what had been said at the conference about the nature of color vision. At another level, I would be speaking to a great mixture of people who knew nothing of our ongoing conversation—men and women from the neighborhood, bridging Columbia University and Morningside Drive on the one hand and the beginnings of Harlem on the other; children in the choir; tourists staying at downtown hotels; connoisseurs who sought out the cathedral for its architecture and its music, or who had discovered St. John's as one of those few places that maintain the medieval image of a cathedral as a center of intellectual and artistic excitement.

For most of the congregation, my remarks would have to stand without the context of the conference, required by courtesy to fit into the service and the genre of a sermon, but I was determined to address my colleagues as well. It was a daunting assignment, but after all many human encounters are experienced in different ways by different participants, like the very best of children's classics, read aloud by parents enjoying a different layer of humor. The Anglican communion has always included a miscellany of saints, sharing the words of a common liturgy but bringing with them a diversity of beliefs. There have been periods, there still are parishes, where uniformity of belief or social class is sought after, but in most places diversity of experience is welcome and printed programs simplify participation for visitors and wandering anthropologists.

Speaking from the pulpit of a vast Gothic cathedral is very different from speaking in a lecture hall or a classroom. You enter robed and in procession. Bearers walk ahead with a cross and huge torch candles. Organ music and incense fill the space. All the details of the service, even the products of the most modern wave of reforms, have been smoothed to ceremonial solemnity. The preacher at St. John's climbs up a dark little stair at the back of the pulpit and emerges alone at the top to see not an audience but the epiphany of blue, a huge blue "rose" window at the back end of the church, a celestial firmament. Every time I have entered that pulpit, my mouth has gone dry, and I have clutched my notes and wondered what I could possibly be doing there. When I recollect myself and begin to speak, the echo of every word rolls back over the next one, as if each sentence had to merit layers and layers of repetition and long pauses in between. Oddly, I have little temptation in that setting to be didactic, little sense of authority. Instead I find myself speaking with a greater degree of intimacy than anywhere else.

As I approached the topic of perception, I knew that my problem was simpler than that of the princess in the story, for I needed only ask the people sitting in the nave to turn around (this is itself an unusual experience, for children are often told it is improper to turn around in church) and to see the blue of that time and place, a different blue from the bright tiled domes of mosques in Iran but equally the work of worshipful human hands. What does it mean to take that experience and say that it is a complex matter of brain cells and nerve endings leading to an image constructed in the brain? We are all trained to sprinkle sentences like these with disparaging adverbs: only, just, no more than. What you see there is not an epiphany, we might say, it is no more than an illusion, a mere construction of brain chemistry. I thought I saw the eye of God, I might say, peering down on the sanctuary, but it was only a trick of iconography. This is one kind of loss. There is a different kind of loss in saying that the experience may be different for every individual in the church. But it could still be true to say it is the eye of God.

The descriptions are all true. Taken together they offer enrichment, not loss. The problem lies in those modifiers that seem to diminish the experience. In fact they make it more important. First, in the revelation of the intricacy of the human mind, how wonderfully we are made, so that the artifacts of glaziers and the artifacts of neurons are joined to—rather than contrasted with—the wonders of nature. Then, in the realization that what we see is there only by virtue of our individual seeing. It can never be true that "every prospect pleases and only man is vile," for it is the human eye that responds to each pleasing prospect, that knows and creates the pleasures we call blue, yellow, red, green from the physical phenomena. In the congregation children and adults, women and men, from the immense polyglot crowd that moves through New York City, are all constructing their visions in the intricate flesh of their brains, none of them exactly the same.

We have no way of knowing whether any comparable experience of beauty exists for any other organism, although we can test the sensory range of other animals and know that it varies from species to species. We can wonder at the constructions of the bowerbird, but we cannot admire them as does his mate. We can match the colors of a coral reef with those of the fish swimming above it, and know they belong to a single composition, but our delight is private. Like other species, we have our own genetically determined blindnesses, captivity within our portion of the spectrum, unlike insects and snakes, whose sensory systems escape at either end. Many errors must arise from phylogenetic limits on perception, but even more from those we have learned to impose.

It is true that human actions dirty the sky and the rivers, but human vision creates unique versions of their beauty. This in turn represents a responsibility, for the visions in our minds are transient. It is all too easy to forget what the woods were like when they were full of birds. Even as we try to reduce the material pollution of the air, we need to retain untarnished the vision that makes the sky beautiful. If you sit with a child, looking at the sky, you can propose the endless possibilities in the clouds, saying, Look, I see an old woman, I see a rabbit, I see a tree. A child will respond with visions you have not seen; children walk among strangers and find new Edens among the trees and animals and people of everyday life.

The responsive eye enlivens the vista to which it responds; the play of imagination is one way of enriching and conserving the natural world. There is a story told about J. B. S. Haldane, the great British evolutionist, who was asked what, on the basis of his knowledge of the creation, he could infer about the mind of the Creator. His answer was perhaps a joke but surely a revealing one, "an inordinate fondness for beetles," he said. The patterns on the carapaces of the earth's multitude of beetles, thousands of species still undescribed and many threatened with extinction, are also epiphanies. A Swiss painter, Cornelia Hesse-Honegger, moved from painting their blazonry to the discovery of unrecognized genetic damage near nuclear plants. The eye of compassion is as rare and valuable as the beings for which that compassion is felt. Its sensitivities depend on picking out one pattern from the mass and recognizing a kinship to it. To conserve and focus compassion, we often depend on single images, a poster child who represents a whole generation devastated, the spotted owl who stands for the preservation of an entire ecosystem.

Like the design of a Persian garden or a carpet, the stained glass mandala reflects a cosmology, a theory of the universe, which it asserts. I could state that theory of the universe without accepting its complex elegance, but I remain moved and more than half persuaded by the artistry in glass. Its circularity argues a theory of perfection that will affect my day-to-day living and my attitude toward dying; its color suggests a hierarchy of values under heaven; its silence more than half-persuades me that the organ is playing the music of the spheres. Had I lived in the time when its prototypes were built, I would have found its argument overwhelming, for in those days most interiors were dark and unfinished, candles were expensive, and the drama of the liturgy put every other performance to shame.

We have been taught to see in new ways because of the distances we travel casually and the extraordinary sights we see on television and movie screens. A medieval peasant would have seen this window as awesome. A modern adolescent can see it and name it as "awesome" and mean something quite different. It is worth studying the attenuation of the dramas and authority of the church that has occurred over the last two or three centuries, for it prefigures a similar attenuation of the authority of education, for many of the same reasons, and government may be next.

Both interior and exterior landscapes can be made banal, drained of significance. Yet vision can be enriched as well as impoverished. Often symbols are infused with new meaning. We have the possibility of borrowing the stained-glass mandala to refer to the living planet of which we are part, setting it beside the actual photograph of the earth taken from space, making the medieval fascination with circularity a symbol of ecology. No other being that we know of, no generation in our history, is capable of juxtaposing these images or imagining that analogy. We live today with multiple representations, some we call science and some we call art, precise, abstract, vivid, and evocative, each one proposing new connections. I can hold the knowledge of these immensities in mind as I lift and hold the blue and green ball on my writing table, the "something blue" of this time and place.

Ways of understanding are integrated works of art created by many minds, like cathedrals, as much masterpieces of the human spirit as the Greek tragedies or the paintings of the Renaissance. Human beings construct meaning as spiders make webs—or as appropriate enzymes make proteins. This is how we survive, our primary evolutionary business. We differ from other species in that clusters of human beings have constructed alternative visions to be passed on, often reshaping them in the passing. We live, more than any previous generation, in an era where these visions meet, each potentially compensating for the blind spots of the other. If we can find ways of responding as individuals to multiple patterns of meaning, enriching rather than displacing those traditional to any one group, this can make a momentous difference to the well-being of individuals and the fate of the earth. What would it be like to have not only color vision but culture vision, the ability to see the multiple worlds of others?

With the instruments and findings of science we can refine a given pattern of perception, but the mental imposition of a pattern of meaning is the only way to encounter the world. Without it we are effectively blind. We move through metaphors and analogies, learning through mistakes. We necessarily learn in every generation to put our faith in a flat and solid earth, learned with the first steps, even as we unlearn the containing sphere of the womb; then we learn that the flatness too is an artifact of perception. The rhetoric of merely, dismissing these makeshifts of understanding, suggests that there is some stripped-down way of knowing the world, but there is no alternative starting place. In the end, it is possible to combine the sense of being a part of a larger whole with the sense of riding a sphere through space, to combine the assurance of inclusion with solitude.

It is not only in memories of childhood that moments of alternative perception can be found; they lie curled within the carefully rationalized, domesticated patterns of adult seeing that are so heavily overlaid with cultural patterns. Yet as adults, if we have not learned socially acceptable ways of classifying the diversity of experience, we often dismiss the possibility of more than one way of seeing. Reasonable people can see things differently. "Reasonable people" also see things in more than one way. Quite ordinary daily experience may contain within it a multiplicity of vision that proposes the willingness to question what has been taken for granted. Certainties fluctuate over time, so that a respect for one's own past variability can open the door to accepting a multiplicity of views in the present, worthy of respect in oneself as well as in others. Everyone has been both in and out of love.

I saw a vision once by a stream in New Hampshire. Light filtered downward through the trees was echoed by light reflected from the moving current, dapple meeting ripple. A multitude of leaf and fern shapes, fully etched, spoke clarity and presence. Believing in God, I could feel the presence of God. Not believing, I could tell myself some other story of the sacredness in the wood and my sudden recognition of it. But the lack of a name might make me turn away from the undomesticated moment, scratch it from memory. Such moments, labeled from a shared mythic tradition, become the basis of certainty; unlabeled, they may be too disturbing to retain. The lack of a theology can make one tone-deaf to the event, but many theologies could serve. Pan, quietly playing his pipes off in the distance. A naiad, rising from the waters of the brook. A wingbeat of the Shekhina or the Paraclete, or the stirring of a Jungian archetype. Any one of these can give license to retain and treasure the moment, but each entails a set of additional beliefs and assumptions. In Western culture, we all too easily see natural phenomena as inert and as commodities, surely a very peculiar kind of vision.

A name makes it easier to say yes to the moment. To say, This beauty is called blue. Different names for experience that cannot be shared or confirmed become matters of contention and even warfare, while the lack of a name opens the door to the anxieties of hallucination, error, insanity. Yet artists can convey that sense of presence without evoking any of the names that divide us. The newly in love do see the beloved on street corners and subway platforms. A patient in therapy may recognize the face of the therapist in a poster or on a frieze. The newly bereaved do hear voices or, watching by the body, see the chest rise and fall. Declared in poetry, none of these experiences is threatening, but in prose they make us worry about our friends' mental health or our own. It is easy to reject the knowledge of one's own mysterious changeability, like those who rationalize away intense love when it is followed by disappointment.

Surely everyone has moments, often unlabeled, often very brief, when perception is somehow changed. Most people learn to forget them, as they learn to forget or reinterpret the experiences of childhood, to label them as dreams or segregate them with theological or psychiatric labels, protecting themselves from awareness that the world they have learned to see is a depleted one. A few people escape these limits, learning to use their unlicensed visions and even to offer them to others in the arts. A few pursue the images of possibility into new technologies. Others rush from such experiences to dogmatic commitments that close the door to future variation in perception.

That stray impression of a vision in the woods, with all its vividness, may have simple physiological or psychological explanations. It may have fit some template or evoked some earlier learning, perhaps even the first encounter with light on emergence from the womb. Joan and Erik Erikson have pointed out how often the images of deity match a prior image of the nurturing mother, the tender face and focused eyes seen from below. But vision is more than its mechanism, whatever the mechanism. Insight, as some mystical traditions teach, involves affirming both, "This also is Thou," and "Neither is this Thou." Such moments offer the awareness of other levels of perception, other ways of being, a continuing awareness that the beloved lives, the forest and the stream live. It is an impoverished life that makes do with single vision.

Most of the time we are orthodox in our cultivated unawareness of our bodies, but even a neurologist can be shaken by the realization of how deeply perception is shaped by forces of memory, desire, or expectation, other than the messages of the senses. Oliver Sacks wrote of an episode in which his brain simply lost the knowledge of an injured leg— the opposite of the experience of amputees haunted by phantom limbs—and had to learn it anew. One of the common experiences of LSD, they say, is the experience of looking at one's own hand and seeing it as if for the first time, different, transparent, the blood moving in the veins. But hands, however wonderful, must be kept in use to work, scratch, caress—only occasionally, each time newly marked by age and labor, to be signs of revelation. We need both the numinous and the practical. We need to be aware of our own hands in more than one way, to cultivate the fruitful vision that repeatedly finds wonder in hand or stream or cloud.

Sometimes when I talk with friends who spend hours in formal meditation it strikes me that they are seeking therapy for a wounded capacity to attend. As a society, we have become so addicted to entertainment that we have buried the capacity for awed experience of the ordinary. Perhaps the sense of the sacred is more threatened by learned patterns of boredom than it is by blasphemies.

In our society, some experience altered states of consciousness through hypnotism or drugs or breathing techniques, but using these mechanisms also makes it possible to dismiss the experience by attributing it to an external force. Still, the actual content of a hallucination, the symbols and images it works with, must come from somewhere, and some of the perceptions of altered states of consciousness must be present at other times, but suppressed. Even knowledge gained through experimentation with drugs has a reality.

Other societies have elaborated psychological and physiological ways of altering consciousness and nuanced ways of sorting out and classifying experience to fit the needs of individual and social life. Monastic visions (some of them produced by fasting) could be judged by comparing them with orthodox doctrine, some attributed to divine and others to demonic forces. In the Islamic tradition, mystical poetry is full of the metaphors of erotic passion and of intoxication; these metaphors depend upon genuine similarity. The poetry is a reminder that the oddities of perception—the illusions— that occur in love and inebriation represent added ways of seeing the world.

In attending deeply to children and trying to empathize with them, as in studying other cultures, one is constantly reminded that these beloved strangers are behaving in ways that are only intelligible if their world is recognized as differently structured, laid out according to different landmarks. Much of the time we are busy trying to talk children out of their perceptions, giving them the correct answers, the ones that are widely shared and fit neatly into familiar systems of interpretation. The fable about the princess who saw blue can serve as a reminder of the negative potential of the encounter between the generations, the talents that are suffocated, the imposition of particular kinds of societal blindness. It takes adult effort to turn bright, open children into a sullen underclass or into compliant factory workers, to keep life in shades of black and white and avoid new learning.

I doubt that I would have invented the fable of the princess who saw blue if I had not suspected that, for all their power and cleverness, there were things I could see and know that the grown-ups could not. But the larger plot of this chapter is one in which I as an adult am learning from my own memories of childhood and using them to communicate more widely, feeling my way toward a sense of the world that includes both the empirical and the intuitive. Learning, we build on existing knowledge. It often becomes necessary to find a prior wisdom, an earlier layer of learning, to strip off some distorting overlay and combine the recent with the old in Fibonacci ratios of awareness.

In the end, for the fable to make sense, the awareness of color must work its way through the entire fabric of experience. The webs of perception and meaning that human beings construct tend toward integration. What does not fit is likely to remain invisible, unnamed, unattended to. Colors are not known separately but as part of a system of discriminations, yet such is our wealth that I can go home from the cathedral to leaf through catalogs that offer me clothing labeled "navy" and "periwinkle" and "azure." In some languages, green is defined by blue on the one hand and yellow on the other, while in others it is part of a compound of all that is fresh and cool and moist, contrasted with the heat and desiccation that lie along the spectrum from yellow to red. Judging by the recurrent elegance of grammatical systems, the search for organization is the normal state of the human mind, asserting itself in classifications of colors as much as in theologies, so resting content with trivial pursuits may be a symptom of pathology. Still, the same capacity that is used to make sense of ambiguous cues can become rigid to exclude them as meaningless or unacceptable.

Memory permits a cross-cultural journey into strange ways of life that were once our own. It has become commonplace to speak of the discomfort that occurs on meeting members of another culture as "culture shock," but the same experience can occur in rediscoveries of the self. True culture shock occurs when differences run deep and immersion is complete, so much so that ordinary assumptions are overthrown, when panic overcomes irritation. In severe culture shock, one may feel that one is going insane. Yet everyone has traveled to wondrous places, and most of us lack the words to tell the tale or even to remember it.

Other worlds contain fabulous monsters. Deities. Parallel spheres of being. To understand another culture, one must include ghosts and spirits in an explanatory system, as well as the abstractions—pride, honor, sin—that appear as reasons for action. The great ideas of human history do not, after all, refer to anything visible to the eye. When we try to translate from one language to another and from one system of categories to another we discover that categories slip and slide, never matching perfectly. We make the same discovery in the encounter with children and lovers, with the living landscape and even with the layered self.

5

A Mutable Self

AFTER I DECIDED to use the old rhyme about what a bride should have at her wedding, I looked it up and discovered among its multiple versions a third line I had never heard. "Something old, something new, . . . And a silver sixpence in her shoe." What the tradition no doubt had in mind was prosperity for the new household, but the line reminds me of the times when I tucked a five-dollar bill into my shoe or my bra, "mad money" that would allow me to get home on my own if a date went sour.

Today, even if they choose not to be employed after marriage or while their children are small, young women are well-advised to be able to support themselves and to maintain that capacity by use: getting out, working with others, being effective. Circulating. That silver sixpence evokes a cultural expectation for healthy development of boys that is becoming more important for girls: a sense of self that is autonomous, independent, self-confident.

A silver sixpence is hard, round, stable. No fuzzy edges and little apparent change over time. It is a bad metaphor for the self but useful to show how the self is sometimes regarded, a thing rather than a process. Like the blue sky, the self is a matter of understanding and experience. Like money, it is a matter of convention. Think, then, of a sixpence in an inflationary economy, its value based only on agreement and steadily running down unless it is kept moving: earned and spent, invested and combined.

American culture has gone further than most in valuing the autonomous self, downplaying the importance of relationship. It was once virtually unique, for instance, in the preference for having infants sleep alone. Through history, most human infants have slept in the same room, often in the same bed, with at least one adult, then slept with siblings as they grew older. In Manila you can see the carryover of this as people cluster together, pack themselves cheerfully into tiny spaces, and walk, men with men and women with women, holding hands, linked and enfolded, seeking contact. The Jesuit scholastics used to scandalize their American mentors by cheerfully sitting on each other's laps. It was not easy for American Jesuits, moving like armored vessels, their compasses set in the individualistic spirituality of the Counter-Reformation, to train these aspirants in separateness.

Only poverty, it is sometimes implied, would make parents share their bedroom or their bed. This once near-universal human closeness is now seen as bad for the child. New mothers are often told severely that if they bring their infants into bed with them they may suffocate them; sometimes that pressure is reinforced with innuendo about sexual abuse. We have so little information about forbidden sexual activity within households in other cultures that it is hard to be certain, but it strikes me that some violations of the incest taboo depend on distance, on a deficit of intimacy. The very close contact in which kibbutz children are raised, like a group of siblings in the children's house, seems to lead to a reduction in sexual interest in each other, so they grow up and seek romance outside the kibbutz.

Putting the baby in a separate room must lead to a heightened awareness of separateness and a more palpable time lapse between need and satisfaction, especially when an infant is left to cry, tasting abandonment as often as freedom, a self rooted in solitude. We think of this as laying the groundwork for independence, yet times of sleep are imposed and enforced by adults, not chosen. For one held in a warm bed between two protecting bodies, combining the scents of different skins and different rhythms of breathing, it must be easy to feel oneself part of a larger whole. Lying in the sun on a hilltop, you can have the same feeling of immersion in the living, breathing biosphere, full of scent and rhythm, as if you had never been expelled from the womb.

Becky and Shahnaz offer a glimpse of alternative approaches. Becky maintained a visual connection with her mother, but she will grow up to face the possibility and the compulsion of autonomy and will need to be able to make herself at home in strange places. Shahnaz acted almost as if she and her mother were one body, yet she was acutely aware of boundaries beyond them, for the sense of being part of a larger whole may go beyond the family to a community, yet may set that community at odds with other human groups. Iranian boys also separate from their parents more gradually than American boys and may suffer from deep loneliness if they travel abroad; decisions of career and education are often made for the welfare of the extended family. In Iran, not only do infants often sleep with their mothers but they stay by their mothers, dozing and waking, until the mothers are ready to sleep themselves. The reality of human infancy is dependency, but some cultures create charades of early independence and project individualism and the rhetoric of rights onto infants, sometimes even onto fetuses. Other cultures go to the opposite extreme: autonomous membership in the wider community may be conferred only at adolescence or later, perhaps never for females.

It was in the Philippines, where I became pregnant myself, that I began to think of personhood, both the inner sense of self and the assurance of membership, as something that comes into being and grows through relationship and participation. I wrote a poem in that period using the imagined image of the Virgin Mary to express the way mothers and other caring adults turn infants into members of the human community:

In cradling that small god she had conceived,
She made him Man by loving.
Mothers do.

Not only mothers, and not only in infancy. The gift of personhood is potentially present in every human interaction, every time we touch or speak or call one another by name, yet denial can be very subtle too, inflicted in the failure to listen, to empathize, to attend. Membership in a human family or community is an artifact, something that has to be made, not a biological given. Membership both acknowledges and bridges separateness, for it is constructed across a gap of mutual incomprehension, depending always on the willingness to join in and be changed by a common dance. Western culture associates independence and autonomy with strength, but there is a sense in which an awareness of being part of a larger whole, of being defined by context, a self in adaptation, can offer a different strength, leading to flexibility and constant learning. One can define a human being by DNA or by the physical traits of the species, but I prefer to use the word *person* for the focus of a pattern of relationships. Caring and commitment are what make persons, and persons in turn reach out for community. Personhood arises from a long process of welcoming closeness and continues to grow and require nourishment over a lifetime of participation.

A willingness to offer full participation to all its people is in some sense the criterion of a good society, yet societies vary in their ideas of where membership begins and how it comes into being. As I write, debates about the rights of the unborn and of those impaired beyond all capacity for participation are ebbing and flowing in different places in the world. The recognition that personhood is socially constructed means that there is no single, self-evident answer to these debates. Almost everywhere, however, a person is one who knows others even as she or he is known: more than living tissue, a participant. Exclusion and second-class membership, when full humanity is denied, are assaults, bloodless murders. Just as the creatively responsive eye implies responsibility to preserve natural beauty, the fact that personhood is culturally constructed increases responsibility rather than decreasing it. There is no alternative method.

Not only is a fetus contingent, a part of a woman's body,

but an adult, man or woman, is also contingent, part of a larger whole, family or community or ecosystem. We cannot afford to carry emphasis on the individual too far, for no one—fetus, child, or adult—is independent of the actions and imaginations of others. Persons are human individuals shaped and succored by the reality of interdependence.

Those who see personhood as coming into being at a single point in time, whether through a divine act or through the biological events of conception or birth, uphold a lonely vision of the self rather than the self in relationship. Such an absolute vision is likely also to be static, playing down (and often subverting) ongoing development and learning. The self in relationship is necessarily fluid, held in a vessel of many strands, like the baskets closely woven by some Native American tribes, caulked tightly enough to hold water. The possibility of being freely welcomed and cherished may seem trivial compared with life and death, but emphasizing birth rather than participation leads to a society that supports the life of patients in irreversible coma yet denies adequate education and health care to vast numbers of children.

The business of human community includes the shared construction and conservation of meaning and compassion that exist only as they are lived. No legal definition can free us from the need to bring one another into being. I am only real and only have value as long as you are real and have value. What would happen if we learned to read Descartes's *cogito* beyond the first person concealed in the Latin verb forms: "You think therefore I am. I think therefore you are. We think . . . " Every *ergo* conceals a different theory of the intermingling of lives. In Pilipino, there are two forms of *we*, one including and the other excluding the listener. It may be that the more closely one is defined by membership in a group, the more difficult it is to recognize the personhood of the "other" coming from outside.

During the months when the revolution was building up in Iran, I was on the Caspian coast at the campus of a new university. We were set down in a small community where the Iranian faculty were also outsiders, newcomers from the big city; feeling vulnerable themselves, they feared that for-

eigners would act as lightning rods and asked us not to circulate in the town and to be as inconspicuous as possible. That
request had to be honored, but I found that over time seclu-
sion had an insidious effect on my morale, showing me what
is meant by referring to many urban environments as dehu-
manizing. Sometimes I began to believe I was necessarily an
object of hostility, susceptible to attack at any moment.
Sometimes I felt invisible, a nonperson, all my effort in
learning Persian canceled, and unable to fend for myself.
Barkev had teased me when we first traveled to the Middle
East about the fact that blue eyes are regarded as bad luck,
saying that women snatched their babies away when they
saw me coming, and now that joke felt true.

On visits to Tehran, however, no loyalty to colleagues
held me back. There I went and walked among the crowds.
Sometimes a passerby would make a hostile comment, but
when I responded without contention in Persian the tone
would change. As I walked along a main street, full of
demonstrators surging out of the way as armored cars firing
blanks rolled back and forth, people would draw me into
storefronts, taking my hand or my arm: step back, be careful.
These encounters on the street were an acknowledgment of
me as a person and outweighed the risks, which were really
rather minor. If this wandering looked like courage, it was a
courage arising from a contingent sense of self, not an invul-
nerable one, for to me separation was the greater danger. If it
looked like folly, it was folly arising from need.

Within the framework of Western assumptions, we begin
to know a little about how the self is differentiated from others, how it takes shape for males and females, the kind of
resilience associated with it. A wide range of pathologies
have been associated with flawed attitudes toward the self:
lack of self-esteem on the one hand and narcissism on the
other. Physical violence and sexual abuse deform the sense of
self, or split it into multiples. So do insult and bigotry. So
does invisibility or the realization that in a given context one
is inaudible. We think of the self as a central continuity, yet
recognizing that the self is not identical through time is a
first step in celebrating it as fluid and variable, shaped and
reshaped by learning.

In Iran, looking for a school for Vanni, I visited a kindergarten. The teacher announced that it was time to draw and walked around the room with a cardboard box from which she gave one colored pencil (no opportunity to choose) and one piece of paper to each five-year-old. After a while she announced that drawing time was over and walked around the room again, this time carrying a wastebasket into which she put the drawings. Coming from the land of the decorated refrigerator door, where we have been taught to applaud each child's efforts as a way of building self-esteem and independence, I was appalled.

Respect for children as individuals and support for their emerging creativity continued to be the criteria for my selection. At the same time, I reflected that Western ideas of the individual are not universal and that American styles of child rearing are not noted for promoting cooperation and sensitivity to others. Particularly for boys, we value separateness: separateness from family, from community, and from the natural world, which we feel free to dominate and exploit. Even though girls are expected to retain a sense of connectedness, they are disparaged for it, while all too many boys are pushed into proving themselves by aggression and competition.

To find really profound differences in concepts of the self or the individual, I would have to look beyond the cultures in which I have worked; for Judaism, Christianity, and Islam are similar in supposing a self separate from God, free to choose obedience or not. I would have to look instead at hunting-gathering cultures or at the highly elaborated psychologies of Buddhism or Hinduism. Even so, there was enough variation in my experience to make me aware of differences. Israeli teenagers learning mutual support in the desert had first taught me something about different constructions of the self, and that in turn allowed me to understand what I was hearing when American brides complained of the intrusiveness of Iranian in-laws treating their time and property as common to the household.

Because the self is the instrument of knowledge, different concepts of self offer different criteria for truth, whether social or private. Authenticity and sincerity are not private but interpersonal, with very different meanings in different cultures. Like the concept of zero in mathematics, a concept of self is pivotal in organizing experience, useful as an idea as long as it is not mistaken for a thing. Yet even though we regard the self as logically central to any way of experiencing the world, we are trained to look through it like a pane of glass, only noticing when it becomes blurred or cracked. The Western insistence on a separate self carries its own blindness, its own nonrecognition of necessary connection, its own inconsistencies. The very self we set out to affirm can become a hostage to fortune.

The self is learned, yet ironically it often becomes a barrier to learning. The illusion of autonomy confers a sort of immunity, often tenaciously defended, to the effect of new contexts and relationships, yet in order to move through society, we are asked to put the tenuous certainties of the self at risk again and again. The self fluctuates through a lifetime and even through the day, altered from without by changing relationships and from within by spiritual and even biochemical changes, such as those of adolescence and menopause and old age. Yet the self is the basic thread with which we bind time into a single narrative. We improvise and struggle to respond in unpredictable and unfamiliar contexts, learning new skills and transmuting discomfort and bewilderment into valuable information about difference—even, at the same time, becoming someone different. Clarity about the self dims and brightens like a lamp in a thunderstorm or a radio signal from far away, but all our learning and adapting is devoted to keeping it alight.

The cost of exposure to another system of certainties is a bruising risk to clarity about the self. Taking on a new role or entering a new institution are both transitions when the self is put at risk: school systems are often particularly violent in their attack. Thus, children who fail to learn in school may simply be unwilling or unable to put a fragile sense of self or of membership in a group at risk, while adults who decline to learn do so in self-defense.

To become educated, one must concur with the implication of ignorance—and in many traditions one must also concur that one is evil, a sinner. In societies with immigrant communities, many children have to concur that their parents are ignorant, while members of minority communities may get the message that ignorance is their permanent condition. These are very expensive agreements to give. Traditionally the definition of oneself as ignorant has been compensated by the promise that, at the end of some number of years of submission and deference, one will be allowed to become somebody—a pillar of adult society. For the many children who suspect that this promise is false, the bargain is unacceptable. Even in private schools, with their constant message of selectivity, the insults of schooling are barely tolerable. Even when we try to build up the self, we subvert it for the sake of discipline and conformity. It is almost as if schools demanded, Leave your self, your self-esteem, the confidence accrued from learning to walk and speak, at the door. And do that without the genuine confidence that in the end you will have a share in your society and that being an adult is desirable. This is what many children are asked to do in school. I can't think why anyone puts up with it.

When Vanni was approaching the end of secondary school, one of her teachers warned the parents that he would soon be assigning personal essays of the kind that students write to project their personalities and talents for college admissions. Vanni turned to me in some distress, and said, "Mom, I don't know how to write an essay in the first-person singular."

In many schools, children are disciplined from early on not to do what is called personalizing. Not to use the word *I*. Not to give their opinions or use school essays as a vehicle for self-expression even while they continue to be vehicles for competition. What I suggested to Vanni, since her primary interest is acting, was to shift to a different medium, where her capacity for expression had not been deformed by classroom conventions. She developed and taped a dramatic monologue and edited the transcription. But what an extraordinary thing it is that in a society where we regard the self as central, we are so often engaged in silencing its expression or putting confidence at risk. Volumes have been written about the miracle represented by learning to use the word *I*, yet that capacity is under constant attack.

Children learn skills and information in school. These are the issues when we complain that they cannot use decimals or give the dates of the Civil War. More significantly, they learn how society is organized and where they fit into that organization. They learn notions of authority and truth and the limits to creativity. These are the underlying communications of school. For a very large number of children they have been basically negative, a progressive stripping away of dreams, an undermining of confidence. Western societies and their imitators use competition to improve performance, beginning in the classroom, paying a price in the loss of collaborative skills. For every child whose confidence is enhanced, there are half a dozen for whom it will be reduced, and some of those will grow up to inflate their own self-respect by finding someone else to put down.

In the Philippines, children learn early on to avoid competing, so that success or failure is often attributed to luck and the child who claims credit for success is remorselessly teased. When American educators arrived in the Philippines, they found this reluctance to stand out and excel frustrating. American clergy have sometimes complained that Filipinos lack the kind of internalized conscience they were used to, responding to public shame rather than true contrition. Individual behavior is experienced as an expression less of the self than of the group. Yet if the conscience really is an internalization of external authority, as Freudians argue, this may be liberating to know.

Even in a society that uses competition to select and strengthen a few members of the group for success, there are situations in which the smart ones, the successes, limit their risks in the face of future challenges, for once they have gotten away from school and become established at a high level, the risks of learning may seem hardly worth it. Given a choice, as we are later in life, most people choose not to learn and therefore not to change except in superficial ways. Deborah Tannen points out that in American culture men are notoriously unwilling even to stop on the road to ask for directions, but this is only one of many settings in which obtaining information or guidance is blocked because of the acknowledgment of weakness involved. Only a few people become, out of their experience, addicted to the process of learning, to its intrinsic rewards.

When we went to the Philippines, Barkev and his colleagues went as professors of management, brought by the Ford Foundation. They started trying to learn a little Pilipino—phrases and greetings. One day one of them walked into a room where there were six young Filipina secretaries and complained loudly, "No breasts, no breasts!"—a tiny mispronunciation of "no keys." The room was filled with cascading female laughter, and the story was retold for months. That was the end of their effort to learn the language, for the professors were prohibited by their status from making fools of themselves. Wealth and power are obstacles to learning. People who don't wear shoes learn the languages of people who do, not vice versa. Given a choice, few will choose the reversal of status that is involved in being ignorant and being a learner, unless there is a significant gain of intimacy or respect in the new learning. When I first became a dean, I admired the campus skating rink and started talking about learning to skate, but helpful faculty friends argued that as dean I could not afford to let colleagues see me in the inevitable comic falls.

The Jesuits at the Ateneo de Manila were a mixed group of Filipinos and Americans who had been there for many years, for when the United States took over the Philippines from Spain the Vatican moved to disassociate Philippine Catholicism from the Spanish tradition and Spanish Jesuits were replaced by members of the New York Province of the Society of Jesus. By the time I was there, a new Philippine Province of the Society had been created, and many American Jesuits who had worked there all their lives had been transferred officially to the new province governed by Filipino priests they had trained. Thus, the Ateneo de Manila, which like every educational institution was a place where the self is battered and buffeted and reshaped, became a doubly conflicted force field of ways of constructing the self.

The Jesuit tradition, shaped by the Counter-Reformation, is based on a training system that deconstructs the self and reconstructs it on new lines. The spiritual and emotional technologies that are today called therapies find their roots in the spiritual disciplines and practices of earlier eras. Typically for a Jesuit, childhood faith is lost and a new, self-conscious faith, both more individual and more individualistic, forged. The Society of Jesus has produced many of the great soldiers of faith and many of the great apostates, including Fidel Castro. Jesuits who survive that formation are tough.

The American Jesuits in Manila, still in 1966 holding almost all the positions of authority in the university, had a certain massive quality of assurance, striding across the campus in their tropical white soutanes. Many had come as young priests twenty or thirty years before; told when they came that they didn't need to learn the language, they hadn't. But attitudes had started shifting by the sixties, and nationalism was increasing. I persuaded the president and the vice president of the university, the dean and the treasurer, and one or two others—all American Jesuits—that if they could, as their rule provided, spend one hour a day in meditation, they could also spend one hour a day learning the language of the people they had to work with.

Certainly there was moral blackmail involved. But I think the main reason I had a measure of success was that instead of turning my "students" into beginners, incompetent newcomers, I tailored a program to acknowledge that they were veterans. I decided that since they already recognized a considerable miscellany of Pilipino used in daily life, my task was to reveal their existing knowledge, help them privately to use it more flexibly, and add to it gradually in ways that would be immediately useful and rewarding. So I made an inventory of knowledge already in place: common greetings, names of songs and cities, the phrases that any expatriate community drifts into using.

Few societies have made the assumption demanded by changing technologies and longer lives that adult human beings are in some significant sense constantly recycling themselves—learning new things and finding a way to balance the pains and the joys of learning—so few societies reward those who take the risks of new learning. Most of us have come out of school rather pleased not to accept the position of ignorance any longer. But the Jesuits had to reconcile that preference with a commitment to humility no longer widespread in Western culture.

Women are often constrained to make new beginnings because of choices made by men, sometimes moving them to new places and cultures. I was in Manila because, after I got my doctorate in Arabic linguistics, my husband accepted a job there. How do you survive under these circumstances? One way to survive is to learn, accepting the internal change that new learning requires and the loss of status that goes with being a beginner once again. In a new country this may mean returning to the infant's task of learning a whole new language and culture, so it is not surprising that many of those assigned to work overseas take refuge in expatriate enclaves and continue to assume that their way of doing things is right, with few changes. Some American women in Manila took courses in Chinese cooking and learned how to judge the quality of pearls, and in Tehran they took seminars on Persian carpets—skills as souvenirs. Husbands are no more willing than wives to risk the changes that would go with fundamental new learning.

Most of the jobs that take Americans overseas are structured so they are not obliged to learn the local culture; indeed they may carry a sort of obligation to hold on to American ways of doing things and to the authority this implies. Barkev has taught management in several countries, in English. To do this, he has eventually had to learn a great deal about the business climate and how people function. He is being paid for expertise, however, not for a willingness to put himself back in nursery school and try to learn the system from scratch. Even in socializing with the local community, expatriates seek out those who will reinforce their sense of confidence and familiarity. Shortly before the Iranian revolution, a member of the diplomatic community said to me, "Look, I socialize with Iranians all the time, and there just is no groundswell of hostility to Americans."

Women have suffered from lower self-esteem than men

and have been less respected and less valued, but the very responsiveness demanded from women can sometimes lead to greater adaptability and greater willingness to follow the cues of a new environment. Today women are especially likely to work for change in cultures where, having been valued primarily for beauty or continuing fertility, they face an earlier loss of status than men, a declining future. There are societies, however, where women's status traditionally rose as their sons grew to maturity. An Iranian woman of fifty, courted and honored by a son coming into his prime, could be poised and confident, with little motivation for new learning or social change, while a new bride, who has recently left her home environment to start from scratch among strangers, is necessarily malleable. Even when he has set up a new household, a businessman may stop off at the end of the workday with flowers for his mother and sit and drink tea with her, listening to her advice, while his wife is at home preparing a meal and looking after young children. Not surprisingly, social change is less attractive to women for whom the best years of their lives are still ahead under the traditional system. In fact, both men and women may be more at peace with the losses that accompany aging in cultures where the old are respected and life has a built-in sense of progression. The Western preoccupation with progress may be an effort to compensate for a personal sense of being condemned to regress.

Many adults only take on the challenge of profound change when they are desperate. This is why so much of adult learning is packaged today as therapy and why it must often offer the compensation of membership in a new community or relationship. We have begun to develop rituals for adults who find themselves in need of drastic change and new beginnings, rituals that give some value to the surrender of adult confidence. Alcoholics Anonymous and other twelve-step groups teach that they cannot help you until you hit bottom and relinquish your sense of being in control of your life—apparently it is helpful to get the acknowledgment of weakness over with in order to make new learning possible.

The alternative would be to conserve the openness and need for new learning that we find in infants, by making it a part of identity. If I were to move to a new country now, I doubt that I would become fluent in yet another non-Western language, because doing so does get harder with the passage of time, but I would learn enough to cope. Learning languages is part of my sense of myself. Following the Turkish proverb that says *her lisan bir insan,* "every language a person," each new language has come to represent an enrichment to me. I know how to do it, I enjoy doing it, and frankly after writing this paragraph I'd be embarrassed not to. I also know that eventually, even though I may look like an outsider, people will recognize their words in my mouth and respond, and that too has become necessary to me.

All too often those who can teach or lead with authority are armored against new learning, while those who are open to new learning are made diffident about expressing what they do know by the very fact that they deem it tentative. The best learners are children, not children segregated in schools but children at play, zestfully busy exploring their own homes, families, neighborhoods, languages, conjuring up possible and impossible worlds of imagination. Only a little way from the front door, in other parts of the city or in forest or meadow, exploration continues to be possible throughout life. Some traditions emphasize this, expecting those who have leisure to fill it with explorations of the arts or natural history. The eighteenth-century idea that a gentleman might collect beetles, read unfamiliar texts of the classics, or conduct experiments played a role in the emergence of modern science.

There is a famous story about two visitors to the Ames experiments in Princeton. Adelbert Ames had set up a series of boxes and rooms (if an artist did it today it would be called an installation) that created optical illusions by distorting perspective. Looking through a slit that allows vision with only one eye, the visitor was invited to touch various points with a stick, but because of deliberately distorted clues of perspective the stick kept missing. Eisenhower, it is said, lost his temper when he visited, threw down the stick, and refused to continue. He had a vein in his high, bald forehead that used to pulse visibly when he was angry or frustrated. Einstein, it is said, was fascinated when he encountered the same errors, using them to explore further.

The two men had clearly found their ways to greatness in the niches that fit their temperaments, but they were also shaped by the conventions of the worlds in which they worked. Generals and presidents are expected to be decisive. An open mind, the willingness to learn from mistakes, the willingness to admit ignorance—these are not widely valued or rewarded in the circles where Eisenhower developed. When political leaders hesitate or revise their views, we mistake it for weakness, not strength. As a society, we need to consider whether those conventions might be altered, whether a little more tolerance for ambiguity might not be a good thing in those who hold leadership positions. We joke about the problem, with buttons that say, "Question authority," and desk plaques that quip back, "My mind is made up, don't trouble me with facts."

There is a paradox here, like the paradox in the Jesuit administrators who were so massively arrogant, teaching with all the authority of the church, and yet were urged to humility by their tradition and by the demands of effective service. It was a balance that only a few seemed to achieve, just enough to argue the possibility. We may not use the word *humility*, but it is becoming important to recognize and value new kinds of fluidity and openness to learning at every stage of the life cycle, in home, school, and workplace.

More flexible boundaries of the self open up attention to the environment that may ultimately be essential to survival, for it is not the individual organism that survives but the organism in the environment that gives it life. We need to find ways to encourage a sense of the self as continuing to develop through responsive interaction. Relying on competition as a way of motivating learning eventually subverts not only cooperation but also the willingness to learn. The models for a more responsive sense of self might be borrowed across lines of culture and gender or be treasured from an undamaged childhood.

Learning is perhaps the only pleasure that might replace increasing consumption as our chosen mode of enriching experience. Someday, the joy of recognizing a pattern in a leaf or the geological strata in a cliff face might replace the satisfactions of new carpeting or more horsepower in an engine, and the chance to learn in the workplace might seem more valuable than increased purchasing power or a move up the organizational chart. Increasing knowledge of the ethology of wolves might someday replace the power savored in destroying them.

We reach for knowledge as an instrument of power, not as an instrument of delight, yet the preoccupation with power ultimately serves ignorance. The political scientist Karl Deutsch defined power as "the ability not to have to learn," which is exemplified by the failure of empathy in a Marie Antoinette or the rejection of computer literacy by an executive. Ironically, in our society both the strongest, those who have already succeeded, and the weakest, those who feel destined for failure, defend themselves against new learning.

Sitting alone at my computer a little after dawn, writing seems a very private thing: my thoughts, my words, the gap between them that I struggle with. But unlike a typewriter, the computer keeps me a part of multiple conversations. A poem from a woman in California, scrolling across the screen, about the inaccessible speech of the body; the machine's curmudgeonly messages, programmed by others, balking at instructions it finds unacceptable—these remind me that I am shaped by other minds. I sit here telling stories about human give and take, repeated encounters sometimes leading to growth, and all the words and concepts I use are old, inherited, part of the way I have been shaped by my environment. I try to become transparent to their possible meaning. The trees on the slope outside grow invisibly and move gently in the wind, shaping me more than I shape them, each one playing a role in birthing a human consciousness. With a sense of self so permeable, peripheral vision is essential, for all those others present with me now are a source of identity and partners in my survival.

Adults are freer than schoolchildren in their writing, but I am in defiance of scientific convention and much of literary history when I claim the freedom to begin many of my sentences with the word *I*. Yet it rescues me from the temptation to be categorical. The word I want is *we*, but there are limits to the assumption of agreement, so I "personalize" as a more honest way to be inclusive. Impersonal writing often claims a timeless authority: this is so. Personal writing affirms relationship, for it includes these implied warnings: this is what I think at this moment, this is what I remember now, continuing to grow and change. This finally is contingent on being understood and responded to.

6

Construing Continuity

IT WAS NOT UNTIL THIRTY YEARS after the senior year
of high school I spent in Israel that I returned in 1988 for an
extended visit, accompanied by Vanni. We traveled around
the country, and I told her stories of what that year had
meant to me, even as I kept trying to understand the changes
in the interval: three wars, expanded territories, and new
waves of immigrants. Vanni's responses were different from
mine at her age, more skeptical. Although she was only a
year older than I had been, she filtered her perceptions
through different experiences, including seven years of childhood in Iran. I was different too: whenever I tried to fathom
how much things had changed, I was confounded by the
problem of knowing whether the change was in myself or in
the country.

In 1956 Israel matched my youth. It offered me, as an
American teenager, a model of commitment that I took away
with me and treasured, so I was startled at the ways individual plans had been redirected, how different my friends' lives
had been from what they had firmly predicted. They still had
the same intensity, but not the same innocent idealism. They
commented constantly on the extent of changes, boasting
and kvetching alternately, but that was a familiar pattern. I
found that I responded to the same individuals I had liked

thirty years before and to the familiar atmosphere of questioning and debate. It reassured me of a basic continuity across the years: that I was in some sense the same person I had been, and so were they.

Israel had also offered me a vision of equality for women, but returning with my perceptions changed by progress and debate in the interval, I was startled to realize that, by 1988, even with a backlash under way, the status of women looked better in the United States—but how to be certain that the status of women had not also changed in Israel, perhaps for the worse, because of increasing numbers of ultra-Orthodox? I was puzzled at things I had failed to see during my first stay and unsure whether I had been too busy or inattentive or simply blind. Why, for instance, had I never climbed up to the fortress of Masada as a teenager? Simply because the archaeological excavations that led to opening the site, where Jews had fought to the death to preserve their tradition, had not been completed. Why had I never been to Yad Vashem, the vast holocaust memorial? I had, but most of it was built later, and preoccupation with the holocaust has intensified since.

I returned again a year later and made a project of seeking out high school classmates from thirty years before, asking them to tell me the stories of their lives in the interval and pondering their experiences of continuity and change. Israel is not an easy site for research, for I found I had to answer five questions for every one I got a response to. Still, I was startled to realize how often, in this context where everyone spoke of change, families hunkered down, couples stayed together, and children settled close to their parents.

Both Iran and the Philippines have gone through revolutions since we lived there, yet continuities keep emerging under superficial change, sometimes a long time later, like the forms of nationalism that emerged in the former Soviet Union after seventy years of Communist rule. Revolutions sweep individuals from positions of power but rarely sweep away the old concepts of power and how to use it, so patterns reassert themselves. The search for change is almost always the assertion of some underlying value that has been there all along, for men and women who set out to build something new bring with them their ideas of what is possible as well as what is seemly and what is comely. Social visions come like brides, dressed in hand-me-down finery.

In Iran after the revolution, old themes resurfaced very rapidly. I had been with the students when they went through the buildings and took down all the portraits of the Pahlavi dynasty at our fledgling university on the Caspian, sharing their euphoria and sense of liberation. The removals left pale patches on the walls, but repainting was not necessary. Within weeks they were covered with portraits of Khomeini on the same scale. One common portrait of the ayatollah standing against the sky was virtually a remake of a favorite picture of the shah. From our first arrival in Iran, the prison at Evin had been pointed out to us as a symbol of the kind of political repression that must be changed, but once emptied it was quickly filled again with dissidents against the new regime. The rumors of corruption among the mullahs and even the members of Khomeini's family sounded suspiciously similar to rumors about the old elite, whether the actual continuity was in a pattern of corruption or a pattern of paranoia about the powerful—or both. The names of streets have changed, the women are veiled, the cabarets are gone, but at a deeper level I suspect that much is the same. The Islamic revolutionaries were seeking continuities with Iran's religious past and, at another level, reasserting an ancient longing for authenticity that recurs through Iranian history. But there is also a recurrence of familiar and unlovely ideas about power and the corrupting nature of social life.

In all learning, one is changed, becoming someone slightly—or profoundly—different; but learning is welcome when it affirms a continuing sense of self. What is learned then becomes a part of that system of self-definition that filters all future perceptions and possibilities of learning. It is only from a sense of continuing truths that we can draw the courage for change, even for the constant, day-to-day changes of growth and aging.

When Vanni was reaching her teens, already committed to a career in acting, she said one day, "Mommy, it must be awfully hard on you and Daddy that I don't want to do any of

the same things you and Daddy do, or Grandma and Grand-
dad." Well, how could I know whether the sense of continuity was critical at that moment or the sense of rebellion?
Somehow both must be present. I crossed my fingers and
said, "You can't be a good actress unless you're an observer
of human behavior and unless you wonder about other people's motivations. Actually, what we do has a lot in common."
I was lucky, since apparently what I said then was useful in
setting up a relationship between continuity and change that
fit her needs. American families have traditionally felt they
were combining continuity and change when the sons of
garage mechanics have become engineers and their sons
have become physicists, but they might as easily have felt
alienation across the differences in income and education.

In Israel I had repeated conversations with older members of kibbutzim bewailing the fact that their children do
not want to "follow in their footsteps," choosing to leave the
kibbutz, even to live abroad. "Did you grow up on a kibbutz?" I would ask. "Oh, no, my father was a shopkeeper in
the city and very religious." The parents had left home to
found the kibbutz, and now the children are following in
their footsteps by leaving. Any social innovation, like the
cooperative living of the kibbutz, is vulnerable to the fact
that the next generation may be more interested in emulat-
ing the novelty of innovation than in continuing the parents'
particular solutions. The pioneers of Israeli kibbutzim
wanted to go back to the soil and to productive labor, but
they tried to retain many aspects of urban life in their attitudes toward ideas and toward the arts, reading, questioning,
and debating political ideals. If the kibbutz movement ever
does fully settle into a pattern of biological recruitment—of
one generation replacing the next—these older stylistic continuities may fade into country ways.

Several years ago I was invited to speak at a national conference of midlife members of a teaching order of nuns. They
all belonged to the age cohort that had entered the order
shortly before aggiornamento in the Catholic Church: when
they entered, they came to live in large, routinized convents,
eating and sleeping and praying on a strict schedule. They

wore old-fashioned black and white habits, with wimples and veils, and they were taught to avoid friendships and personal conversation. Today they live in apartments, some alone and some with other sisters, developing friendships and dressing as they please. As high school and college teachers, they were all well-educated, but today they can choose their assignments, and many work in other social service professions. During the transition from the past to the present, a great many young women left the order—as many as four out of five of those entering in some years— some because of too much change and others because of too little, whether in the order or in the church. The ones I met were forceful women who had been able to embrace a radical change in the pattern of their lives by recognizing continuity in discontinuity. They could assert that their commitment was still evolving, which helped them to be patient with the rigidities of the institution, to find discontinuity in continuity.

Even in our change-emphasizing culture, we often reconstruct and romanticize the past to emphasize continuity, and we retell the lives of great men and women as if their destinies were fixed from childhood. Henry Ford is said to have lost his heart to the first horseless carriage he saw as a child; Teresa of Avila is said to have tried to run away to be martyred by the Moors; mothers of musicians describe their children's response to the radio—you see, these stories suggest, this child was already on the path to this particular kind of greatness. Children are offered models of achievement that minimize discontinuity, proposing a single rising curve of development, marked with a rhythm of recognizable milestones: promise in childhood, preparation in youth, continuing progress in adulthood.

The traditional model of a successful life does not include radical new beginnings halfway through—these, by implication, are only necessary when a life has gotten onto the wrong track—but we do have an alternative story type in which the plot involves a major shift, repudiating a bad course and turning onto a good one. In his *Confessions*, St. Augustine emphasized the wickedness of his early ways, highlighting his conversion, and so did Malcolm X in his *Autobiography*. This way of construing the past is increasing through the development of twelve-step programs, which require that some moment be identified as touching bottom, after which ascent becomes possible. Narratives of discontinuity offer the chance to leave the past behind, the good as well as the bad, yet anyone who claims the liberating experience of being born again must also face again the groping learning of an infant. Some people handle a transition like the disintegration of a marriage by amplifying discontinuity: moving to a new town, growing a beard, getting a makeover, becoming a new person, erasing affection in legal fights. Others minimize the change: "We haven't gotten on for years," they say. "All we did was to make it official."

Often those who have made multiple fresh starts or who have chosen lives with multiple discontinuities are forced by the standard ideas of the shape of a successful career to regard their own lives as unsuccessful. I have had to retool so often I estimate I have had five careers. This does not produce the kind of résumé that we regard as reflecting a successful life, but it is true of more and more people, starting from the beginning again and again. Zigzag people. Learning to transfer experience from one cycle to the next, we only progress like a sailboat tacking into the wind.

A single rising curve is unlikely to reflect the lives of very many in a world where life expectancies approach and then pass seventy. Now the norm of a successful life more often involves repeated new beginnings and new learning. All those who become immigrants and refugees, displaced housewives and foreclosed farmers, workers whose skills are obsolete and entrepreneurs whose businesses are destroyed will have to learn new skills. Increasingly, returning to the classroom and sometimes totally shifting gears from one identity to another will be fundamental to adult development. We will become aware that a zigzag, seen from another angle, may be a rising spiral, so that readjustments are a record not of failure but of growth. Beyond a certain level of economic and technological development, any society must become a learning society, one in which many of the most talented and energetic members have more than one career, as athletes and military people do today. Learning is the new continuity for individuals, innovation the new continuity for business. Each requires a new kind of self-definition.

The traditional tales of achievement or of conversion, whatever suspense and danger were built in along the way, were always cultural constructions, fabricated to make the confusing realities of life fit in with ideas like salvation or progress. People from different cultural traditions see the past differently, whether they are glorifying the industrial revolution or justifying their individual choices, but any cultural theory of the life cycle is likely to leave some people feeling that their own narratives do not measure up. Our narratives are becoming more complicated and ambiguous, and the culturally given plotlines are likely to mislead. The continuities of the future are invisible, horizons in shades of blue we have not learned to name.

In my recent work on the ways women combine commitments to career and family, I have been struck by how commonly women zigzag from stage to stage without a long-term plan, improvising along the way, building the future from "something old and something new." For men and women, résumés full of change show resiliency and creativity, the strength to welcome new learning, yet personnel directors often discriminate against anyone whose résumé does not show a clear progression. Quite a common question in job interviews is "What do you want to be doing in five years?" "Something I cannot now imagine" is not yet a winning answer. Accepting that logic, young people worry about getting "on track," yet their years of experimentation and short-term jobs are becoming longer. If only to offer an alternative, we need to tell the other stories, the stories of shifting identities and interrupted paths, and to celebrate the triumphs of adaptation.

Recently I have been experimenting with asking adults to work with multiple interpretations of their life histories by composing two brief narratives, one focused on continuity, the other on discontinuity. "Everything I have ever done has been heading me for where I am today" is one version of the truth, but most adults can say as well, "It is only after many surprises and choices, interruptions and disappointments, that I have arrived somewhere I could never have anticipated." Most people have a preference, one way or the other, a version that is normally in focus for them, underlying and justifying their current choices, but almost everyone can discover the alternative version. Some solve the problem I have set them by focusing on different aspects of their lives: same spouse, different job; same job, different city. Some notice that the appearance of discontinuity is increased or reduced by the choice of words, so they can make the contrast by saying, "I have always been a writer" and "I used to write poetry, but now I am a journalist." A friend pointed out to me during a period when I was complaining of the discontinuities in my own life that although I had changed my major activity repeatedly, I had always shifted not to something new but to something prefigured peripherally, an earlier minor theme, so that discontinuity was an illusion created by too narrow a focus and continuity came from a diverse fabric and a broader vision.

Continuity and discontinuity are not mutually exclusive. "Wherever I go it's important to me to work with new people." "I have always enjoyed tackling the unknown." "I have always wanted my work to provide new challenges." Some offer metaphors of continuous variation, like surfing, a life of encountering one wave after another. Some say, after years in the same profession or setting, that life is filled with wonder because each day is different. One person wrote, "After all, the laws of physics never change," but, in the words of another, "Sure, I've had the same job for thirty years, but meanwhile consider the turnover in my body's cells." All these approaches can be part of the same repertoire, for these are not exclusive truths. Tales of both continuity and discontinuity can be constructed from the same "facts," the dates and names and addresses of any life history. There is no single true interpretation that must be discovered and held to—on the contrary, each of these interpretations offers a different kind of strength and flexibility for fitting into a society of multiple systems of meaning.

Some dimension of continuity is essential to make

change in other dimensions bearable. The evolution of the Great Atlantic and Pacific Tea Company into the modern A&P is said to have depended on affirming that the company had been and would continue to be a food company. When I was an infant, my mother ruled that it was all right to leave the baby in a strange place with a familiar person, or in a familiar place with a strange person, but too frightening if both were strange. Through most of my childhood, the sense of home was constant, but great numbers of people moved through the household. When Vanni was a child my husband and I moved frequently, but she had a much smaller number of caretakers than I did. Even today, after all those moves, Vanni worries about maintaining friendships across distance—a compensation for disruption? Or a learned appreciation of the value of continuity? Each of us can tell his or her story with alternative emphases.

At some deep level of the personality, perhaps, all change evokes the terrors of abandonment and dissolution, loss of those others who define self, or confrontation with a self become a stranger. During states of high vulnerability, panic is sometimes triggered by minute changes, so we arm ourselves with tokens of continuity. The wanderer or adventurer learns to finger a "lucky piece" along the way, and parents are sometimes advised to encourage an infant facing travel or multiple caretakers to cling to a blanket or a teddy bear, so the same tattered reassurance of the familiar can be carried from place to place. Immigrants and pioneers have always carried with them at least a few objects that provide a link with the past: Grandmother's photograph, a Bible, a pair of Sabbath candlesticks, a small blue china jug that, moved from mantel to mantel, converts a new house into a home. Devout Shiite Muslims carry a little molded block of clay from the site of the great martyrdoms, so that when they prostrate themselves for prayer their foreheads will touch the soil of Qarbala. The mementos I brought with me, reminders of multiple homes, converted an unfamiliar table at the MacDowell Colony into a desk where I could resume the task of composition: recalled in the text, the ammonite and the blue-green globe, the carpet and the Passover plate connect the chapters and lines of reflection.

If I recognize my situation today as comparable to but not the same as my situation yesterday, I can translate yesterday's skills and benefit from yesterday's learning. I will make the mistake neither of trying to start from scratch nor of simply replicating previous patterns. Reinterpretation and translation, so useful in moving from one culture to another, turn out to be essential skills in moving from year to year even in the same setting. But if a situation is construed as totally new and different, earlier learning may be seen as irrelevant. The transfer of learning relies on some recognizable element of continuity—a woman describing her patchwork of careers for me recently remarked wryly on a continuity between work as a kindergarten teacher, a teacher of the deaf, and dean for "Greek life" (fraternities) on a university campus!

Gender stereotypes often suggest that females emphasize continuity (this is called "keeping the home fires burning"), while males venture forth on the new. Yet traditional female roles involve a high degree of unacknowledged adaptation to change, while many men have plied the same trade for a lifetime with little new learning. Women often use labels to construe adaptation as continuity. When farmers lose their farms or men their jobs, their wives may be quicker to adapt, for while the men mourn their homes, their livelihoods, and their identities, the women, equally homeless and impoverished, hold on to their identities as wives, looking for new ways of caring for and supporting their husbands. Men sometimes use labels to assert continuity as well, as when sudden military adventures are justified as defense. The Vietnam War, which triggered profound change both in the United States and in Southeast Asia, was rationalized as a way of maintaining balance.

Along with the odd distortion, now much commented on, of women who used to say, "I don't work, I'm just a housewife," there is a second, usually unremarked distortion when women say, "I've been doing the same thing for the last fifteen years, just looking after the kids and keeping house." If a corporate assignment changes, involving new skills and increased responsibility, the title often changes as well, and the discontinuity (and success in bridging it) is noted. But you rarely hear someone say, "I was getting pretty good at being the mother of an infant, but my new assignment, caring for a toddler, is still really challenging." Mother of one becomes mother of two? Mother of an adolescent? Mother-in-law? My suspicion is that although women may have been less likely to initiate significant change than men, they are highly resilient in finding ways to respond and adapt when change is thrust upon them.

The recognition and celebration of developmental change, season by season and year by year, underline the continuity of family life. Part of the agony of caring for a severely retarded child is the lack of change, the lack of milestones. The definition of any stage of the life cycle only as a plateau, without a dimension of growth, seems likely to lead to stagnation and discontent—living happily ever after is a swamp. The famous "midlife crisis" may be an artifact of such a misdefinition, so too much of the senility observed in the elderly.

Raising children does involve the transmission of continuities, but it also requires sustained and loving attention that welcomes particularity. It involves both providing a base of security and continuing identity, and freeing the individual for risk and experimentation. In a rapidly changing society, parents struggle to make tradition available and to affirm their own continuing convictions while affirming that a child who comes home with a new religion or sexual identity or a Mohawk haircut is still beloved.

The weave of continuity and creativity in the ways that individuals "compose" their lives is not unlike the way they put together sentences and other sequences of behavior. In speaking, we follow culturally transmitted rules of grammar, but these allow totally original utterances; most sentences we speak or hear have never before been spoken, and the most profoundly original insights are only intelligible because they are phrased in recognizable form. Even that family of art forms referred to as improvisatory, such as jazz or epic recitation, actually depend upon endless practice and the recombining of previously learned components so that each performance is both new and practiced. Children need to learn both kinds of skills. No list of appropriate behaviors, no finite set of skills is sufficient.

A story used to be told about the cyberneticist Norbert Wiener. Sometime in the fifties, they say, he was riding in a car driven slowly by a student through narrow streets, when they glancingly struck a child chasing a ball. The student pulled over, helped the child up, crying but unhurt, took her into a nearby pharmacy, got a Band-Aid for her scraped knee and a lollipop, called her mother on the pay phone, delivered her at home a few houses away, and eventually got back in the car with a sigh of relief. Wiener had not moved. "You have hit a little girl before with your car?" he said. The student: "My God, no, heaven forbid." "But then how did you know what to do?" In fact, in order to know what to do in a novel situation, he had to draw on a truly vast amount of existing knowledge: law, psychology, first aid, how to use a telephone. Even in completely new situations response depends on recognizing continuity. Yet from year to year I have to make allowances, as I tell the story, for changes. Neighborhood drugstores disappear, lawsuits multiply, lollipops are branded bad for the teeth and become unwelcome favors.

All around the world we can find transitions under way in which the challenge to leadership is to make change tolerable by providing affirmations of underlying continuity, as Ethiopian Jews arriving in Israel are able to greet the most radical shift in their circumstances by saying they have come home. The nuns who outlasted the reforms were asserting, *Plus ça change, plus c'est la même chose* (the more it changes, the more it's the same).

Without such a sense of underlying continuity, change is so frightening that some are driven into reactionary identities. The pitfall of fundamentalism—whether it is Jewish, Christian, or Islamic, or cropping up in some other tradition—is that when some item is held constant while the context varies, constancy is an illusion, and those who resist change often suffer the reverse, *Plus c'est la même chose, plus ça change* (the more it's the same, the more it changes). The long coats and fur-trimmed hats worn by Hasidic Jews, like the habits of nuns, were only slightly different from general patterns of dress when they were adopted, but freezing these styles created later situations of extreme differentiation. Christian fundamentalists claim that they are practicing "that old time religion," but when they assert the literal truth of ancient words of scripture in the context of modern notions of truth and falsehood, they are in effect asserting something new. Translating the cosmology of the Old Testament into the format of "creation science" turns the insight of an ordered universe into a caricature. The return to tradition that fundamentalists promise, carried out in a new context, often results in radical change, just as, during the Reagan-Bush years, radical change was camouflaged with the label of conservatism.

Sameness and difference are a matter of context and point of view, change and continuity often two sides of the same coin. We can only make sense of the relationship between change and constancy by thinking of them in layers, one flowing under or over or within the other, at different levels of abstraction: superficial change within profound continuity, and superficial continuity within profound change. The deepest changes may take generations, with old attitudes concealed beneath efforts to adapt. My mother once commented that, when a woman who was herself breast-fed shifts to bottle feeding, she still holds her infant as she was held, as if the nourishment were coming from her body; but when her daughter bottle-feeds, the echo is lost. Sometimes the descendants of religious families retain their parents' ethical principles for a generation or more after abandoning the doctrines that supported them. Sometimes groups that have been persecuted act out patterns of self-hate when the persecutor is long gone.

One day a senior colleague told me a joke with a typical mixture of the academic and the mildly salacious, ending with the assertion that physics is like copulation and mathematics like masturbation. He broke off suddenly with an air of contrition and said, "I shouldn't be telling that joke, that's a sexist joke." When I looked puzzled, he explained that it's only for the man that copulation is like physics, whereas, he said, for the woman, "sex doesn't have an object." I was chilled. He had learned that it was a mistake to be overtly sexist and was trying to watch his step, but he had no awareness whatsoever, under the effort at superficial change, of the profoundly sexist nature of the assumptions revealed in his explanation: that sex is something a man does to a woman, a process in which the woman is the object. This was for him self-evident fact.

Changes can be made quickly on the surface that take a generation or more to affect more basic structures, so both progress and degeneration are deceptively slow. In evolution, the very continuity of survival depends upon multiple superficial changes, many of them temporary. If temporary adaptations were to become permanent and genetic (this theory, the inheritance of acquired characteristics, is today associated with Lamarck, but Darwin believed it at one time as well), survival would become harder. Thus, dogs and horses grow thicker fur when exposed to cold, but if a thick coat is inherited, the capacity to adjust fur thickness has been replaced by fixed heavy fur thickness, and the animal can no longer adjust. Natural selection works more slowly, favoring traits that prove useful generation after generation. My father used to point out that a tightrope walker maintains balance by changing the angle of the balancing pole. Freeze the angle of the pole, and the tightrope walker will fall. Death and extinction are the discontinuities avoided by the capacity to change.

The pattern in which corrective feedback brings about the changes or adjustments that will maintain a deeper constancy is what defines a balanced cybernetic system. Human beings can respond more quickly than dogs and horses to messages of heat or cold, putting on or taking off a sweater to maintain a comfortable body temperature. But even among humans the needed adjustment is not available if, for instance, an outer garment is specified by religion. I have often felt sorry for Hasidic Jews in the furred hats and dark coats adopted generations ago in Northern Europe and carried with them to the Mediterranean climate of Israel. In the same way, when educational or political systems are frozen into some form that seems good to one generation, they may lose the flexibility to adapt in the next. Efforts to address social issues by amending the U.S. Constitution instead of learning to read it differently risk compromising the flexibility which is its greatest strength. Protecting the environment with more and more regulations may block the development of improved ways to handle wastes or conserve energy. Rigid standards can undermine thoughtful education, and sometimes overspecific codes of conduct can lead to the atrophy of ethical choice. Only children who are allowed to make mistakes can become responsible adults.

Many kinds of addiction can be seen as efforts to maintain some constancy that at another level brings about damaging change. Withdrawal from a drug is a little like an autoimmune disease, the self-estrangement of a system no longer recognizing itself, and the symptoms of some kinds of withdrawal look curiously like allergies. Alcoholics drink to feel right, to feel like themselves. Sober, they encounter uncomfortable strangers in their own skins, and pouring a drink feels like coming home; their bodies have learned to regard the presence of alcohol as a natural state and adjusted to it.

Chemical addiction is the result of a kind of bodily learning; the learning of ideas also produces a kind of addiction. All views of the world are acquired, and learning a way of seeing the world offers both insight and blindness, usually at the same time. Losing the certainty of a particular worldview can make you feel sick, bewildered, dizzy. From this point of view, culture shock is a withdrawal phenomenon; we reject the new because we have learned to be dependent on the old. In the same way, I may learn to trust someone, premising my life on that trust, and then be unable to reject it when I am betrayed. People will accept martyrdom in order to hold on to an idea.

Sometimes those who have learned to need a particular substance or behavior have learned to need a constant change in the supply, overcoming the adaptive effect of habituation. This need for a constant change develops in some forms of alcohol and heroin use, defining what feels normal. The American economy is addicted to increases in the gross national product, for growth has come to be regarded as the only stable state of a modern economy. Anorexics must get thinner and thinner in order to feel slim; more and more missiles are needed in order to feel secure; pornography must become progressively more horrendous to continue to titillate. Treatments of addiction like methadone or daily attendance at AA meetings are also in a way addictive, but they make sense if they involve replacing an escalating need with a stable one that carries a reduced cost. Perhaps all pleasures that do not bring a natural satiety have an addictive potential. Once money is invented, wealth seems to become addictive, for the wealthy are never rich enough. "Rich" is apparently not a continuity; "richer" is. The present leverages the future.

It is fashionable today to speak of behaviors that used to be regarded as good or evil as addictions: gambling, sex, work, all those behaviors that the person feels compelled to pursue and that may come to seem dysfunctional. No choice, no morality. Yet the moral judgment survives under the clinical veneer in that we do not often speak of couples contentedly and prosperously married as mutually addicted, and we are just beginning to speak of the more respectable substance addictions in those terms. Even those who are proudly free from addictions to morning coffee or afternoon tea are likely to be committed to such learned constancies as jogging or a morning shower or moving their bowels at a particular time, all behaviors that become virtually a part of self-definition by long conditioning. Where is one to draw the line in a thicket of metaphors that both illuminate and confuse? We might do better to see a relationship between addictions and commitments and teach children not so much to avoid addiction as to choose their addictions carefully. Dental floss yes, laxatives no.

The constancies of modern life are increasingly the products of technology: we depend on an information-rich environment, on constant entertainment, on air-conditioning. Some kinds of resistance have been lost, so we have come to require clean drinking water. In a real breakdown of technology, our withdrawal symptoms would kill us. In postwar Iraq, illness and death spread as the society struggled for self-regulation. Some of the constancies we have come to expect, like medical care and increased longevity, are dangerously cumulative. Sometimes, however, a constant in society affects the individual as a violent discontinuity—losing a job is no less painful when economists speak of acceptable levels of unemployment. Sometimes we blame individuals when society is the addict: Who after all really is addicted to crack cocaine? Where is the addiction in the system? What constancies does it maintain?

From the point of view of composing a life or managing an institution, the ability to recognize any situation as representing both continuity and change makes it possible to play that double recognition in tune with changing needs, to avoid the changes that reduce flexibility and the constancies that eat away at the necessities of survival. We know that keeping consumption at familiar levels is eventually going to deplete resources, yet patterns of consumption are oddly difficult to change. When change itself becomes addictive, it seems almost bound to lead to trouble, yet not all acquired habits of constant change are degenerative: constant learning, for instance, is not. A willingness to change in response to a new social environment can be a style of relating to the world throughout a lifetime. Yet the modern vulnerability to boredom may be the long-term result of an addiction to variation.

When I first lived in Israel, less than ten years after independence, military preparedness was a short-term adaptation, but over time, as new generations grew up under arms, it seemed to become intrinsic to the structure of the state. What looked impressive then is worrying today, potentially replacing other constancies of Jewish identity and becoming an end in itself. Thoughtful Israelis today make a distinction between being able to fight for survival, as many of those who died in the holocaust were not, and risking addiction to militarism. There are positive side effects of this drug, for it works to promote solidarity and the maturity of young people—I sometimes wish all my freshman students had spent two years in the military—but there are negative side effects too, such as secrecy and sexism. When I returned to visit Israel, I found I was as concerned about constancies as I was about changes. Yet I could see other processes of change growing into new constancies, like the long-lived forests slowly created of trees planted one by one.

7

Attending a World

THE SOFT SOUND OF RAIN on the roof fades in and out of awareness, along with memories of tropical downpours and the celebration of rain in the desert, not one rain but many. There is a persistent scent of newly cut wood in the room, and the smoky smell of the wood fire, but I only notice these when I come in from outdoors. On my left is a window through which I can look downhill past lichen-covered oaks to a forking stream. How vivid the grays and greens of the lichens are in the rain, the wet bark blackened behind them. Two streams diverge in a . . . I wonder, in a pause between paragraphs, about the many meanings of water, and then how the metaphor of streams would shape our thoughts differently from the metaphor of roads. I muse on the rarity, in the Philippines, of metaphors of binary choice, so common in the West. I check my watch to make sure I don't forget a planned telephone call. Somehow under the ripple of slight distraction, a sentence has taken shape, and I type it into the computer.

It would not be true to say that I am concentrating fully on my writing. My attention is not something I control, not something I fully own, much less a resource from which I might dole out payments. Zen teachers urge students not to struggle against distraction but rather to let the thoughts that come during meditation pass through their awareness, then let them go. When I was in college, I knew a woman who kept fresh apples in her desk drawer so that instead of being restless as she worked she would have the minor distraction of their scent to notice and relinquish. At one time I used smoking the same way, finding a portion of attention easier to focus than the whole.

When I become restless and my thoughts no longer flow to my fingertips, I take my big yellow dog for a long ramble through the wet woods, rebuild the fire, do chores and errands, and then pick up where I left off, to find that my unconscious has made headway in the interval. During most of my life, except for the short periods away in places like the MacDowell Colony, my writing has been fitted in between other kinds of activities. Now this is so much a part of my pattern that when I am guaranteed against interruption I create my own distractions as a counterpoint to the working day.

This is one of many styles of working, a common style for women who have spent years with one ear open for the cry of an awakened child, the knock of someone making a delivery, the smell of burning that warns that a soup left to simmer slowly has somehow boiled dry. My life has forced me to adopt multiple levels of focus, shifting back and forth and embedding one activity within the other, parent and observer, teacher and student. I have been fortunate in living several lives simultaneously, the effect of layers of commitment. There is even room for awareness of the process of learning.

This is an old way of organizing attention for women, not a new one. Women were being pulled in different directions and into patterns of multiple attention long before contemporary conflicts between home and career. They get married and begin to attend to the needs of a husband; then he is jealous of the baby when the baby comes. When the second baby comes, the first baby is jealous of the second baby, the knee baby is jealous of the breast baby, and the husband is jealous of everybody. Women have been trying to balance multiple claims and demands from before the beginning of history, for women's work has always embraced the array of tasks that can be done simultaneously with caring for a child. This has meant taking on whatever could be done with divided or fluctuating attention, could be set down to respond to interruptions and picked up again without disaster. It is not surprising that such work, so easily deferred and juggled, is often treated as having negligible value, yet it must be done and has always included many of the tasks that were central to survival. Women must be one thing to one person and another to another, and must see themselves through multiple eyes and in terms of different roles. Women have had to learn to be attentive to multiple demands, to tolerate frequent interruptions, and to think about more than one thing at a time. This is a pattern of attention that leads to a kind of peripheral vision which, if you limit roles to separate contexts, you may not have. Sometimes this multiplicity can be confusing and painful, but it can also become a source of insight.

Men have been the ones who went out to the hunt or to battle, the first to step voluntarily into unexplored regions and confront alien ways of being. The tasks that have tended to become exclusively male, like hunting and warfare, require sustained concentration, so men have been trained in the importance of single-mindedness, of narrowly focused attention.

Attention involves mobilizing mental capacities to function adaptively in a given situation. Just as situations vary, so do styles of attention. The ability to concentrate exclusively on one thing is essential in the modern world for both men and women, but so is the skill of attending to more than one thing at a time. Ideally, each individual would cultivate a repertoire of styles of attention, appropriate to different situations, and would learn how to embed activities and types of attention one within another. Most drivers are aware of doing this, for instance, for the kind of attention that works best on an open road in the country is wrong for urban traffic, yet both demand focusing straight ahead, and scanning to the sides and the rear, often listening to tapes as well. Over time, all of us develop more than one such pattern, yet in turning from very concrete skills like driving to more metaphorical issues, the discussion of attention often becomes lopsided, emphasizing the need to focus and resist distraction.

The division of labor that requires different kinds of attention from men and women originally made very good biological sense. Among the San people such as the !Kung of the Kalahari (also known as Bushmen), the women are either pregnant or nursing (each child for three years or more) for most of their adult lives, but this does not prevent them from making a full contribution to subsistence tasks. The San used to be called a hunting people, when anthropologists focused almost exclusively on the male half of the societies they studied, but eventually they noticed that some two-thirds of the diet was actually provided by women's gathering and proposed that the San be called a hunting-gathering or a foraging people.

If you are a San hunter, a man, you go out with a small group of companions armed with poison-tipped arrows, hoping to track and wound a large grazing animal, an antelope or a giraffe. (I am writing here in what anthropologists call the "ethnographic present," for the game is mostly gone and this old foraging life is largely a thing of the past.) It would take the poison many hours to work on the animal, but unless you as hunter arrived quickly when it collapsed, other predators or scavengers would get to the meat before you. The San are extraordinarily skilled trackers, able to tell from tiny clues when the animal passed and how it was feeling, taking note of where it urinated, where it defecated, even of where it staggered and brushed against a bush. The veldt speaks to them in great detail, so rich in information that the difference from a city dweller's vision of the same landscape must be like the difference between black and white and color vision. They read the whole condition of the animal they are pursuing from its spoor, as a modern physician might from an array of lab tests. The hunters, focused on the trail, knowing exactly what they are after, must move quickly because the quarry can cover a lot of country before it slows down. When the meat is eventually brought home, it will be shared with everyone in the band, providing a moment of celebration and excitement.

If you are a San woman, you also set off with a small group of companions, but not with the hope of getting one large piece of meat. For a day of gathering, you will probably head to an area where nuts rich in protein, say, or wild melons are known to be found, but in the course of the day you may find and bring home a dozen different foodstuffs: roots, nuts, gourds, melons, edible insects, eggs from a bird's nest, a tortoise, a whole variety of foods spotted along the way.

Because the women move more slowly, they can talk and gossip through the day, whereas the men tracking are moving faster and doing less talking. Several of the women will be carrying nursing infants, and there is probably another child or two along, who refused to be left in someone else's care back at the camp. Some of the children are more or less walking but need periods of rest and will probably ask to be picked up on the trip home, when everyone has the most to carry. In the meantime, the children on foot are zigzagging a little into the bush, so you are watching what each child is doing, looking up in the branches, scanning and checking along the ground for a burrow or a vine that betrays an edible root. Back in the camp women do a second shift, preparing food, looking after children and old people and the sick, carrying firewood and water—doing a variety of overlapping and enfolded tasks. If you imagine yourself as a San woman, you can get the sense of multiple focus that frees the men for the narrower focus of the hunt.

The food that the women bring home is shared without fanfare in the immediate family. Nobody has a party about it—no one gives it that much attention—so it is not surprising that anthropologists took so long to notice its importance. The response to the food gathered by the women is reminiscent of the old notion of a woman as "just a housewife": "not working," not contributing to the GNP, achieving nothing worthy of notice. Nobody celebrates the fruit of her effort. Hunting skills are reserved for men; men do sometimes gather food on the veldt, but they do so unenthusiastically, gathering less efficiently.

With contraception, alternative ways of feeding, and low infant mortality, we no longer need to let the division of labor be determined by reproduction. Men and women do a great variety of tasks, demonstrating a range of potential unexplored in the Kalahari, tasks that require myriad styles of attention. But the advantages that men have enjoyed and the extra value given to their contributions carry over in an extra value given to narrowly focused attention, to doing one thing at a time. The more our society moves toward specialization, the more women and men alike are forced to focus on single activities, living in narrow channels. Yet there are many reasons why less narrow attention, more peripheral vision, offers richer and more responsible living.

My own habits of attention have been shaped by combining different roles. This is useful to me as an anthropologist, for in the field you never know what will prove relevant. Too narrow an attention to the obvious—a dead sheep, say—can make one miss something essential going on at the periphery, or block awareness of the intricate weave of culture. Other professions develop their own special patterns of attention. Employees and junior executives may be trained to attend to one thing at a time, but at the top the peripheral issues can never be fully dismissed, although they can be delegated or tagged for attention when they pass certain thresholds. A leader or chief executive has to include very diverse issues in the range of potential attention—the state of the nation, all the departments and concerns of a corporation or an institution. In politics, as in the tough streets of a city, peripheral vision is essential to survival. Therapists depend on listening and letting their own associations wander, and salespeople learn to recognize out of the corners of their eyes when a purchase might be imminent. Other professions require learned patterns of inattention. Some forms of caretaking require blanking out the awareness of suffering, and waiters often seem to organize their tasks by carefully ignoring customer signals.

In a democracy it is critical to have leaders who do not become preoccupied by any single concern, particularly by the problem of ensuring reelection. The United States has had one recent president, Ronald Reagan, whose attention was severely limited and who delegated too much, and another, Jimmy Carter, who tried to attend too well and in too much detail to matters that should have been delegated. Specialists insist that their preoccupation is always urgent, always more important than any other, very much like small children, but an effective leader must be a generalist who knows what to ignore—and she or he must catch the changed tone of voice that suggests that complaint and demand are no longer routine. The balance is not easily achieved. Bob Cousy, the famous "playmaker" of the Boston Celtics basketball team, was famous for his peripheral vision, for knowing at all times what was happening everywhere on the court, which allowed him to practice a distinctive kind of leadership, quite different from the style that might be learned from, say, baseball or golf. There is a whole genre of anecdotes about the way particular famous people—Napoleon, say, or Churchill—could attend to multiple tasks and conversations at the same time. Chess masters play many games at once, perhaps as their only refuge from loneliness between tournaments.

Even as we compete to receive attention or struggle to know where to give it, it remains the elusive prerequisite of all thought and learning, always selective and always based on some implicit theory of relevance, of connection. Patterns of attention and inattention cluster in every setting and are packaged and pummeled into new forms in school and in the workplace. Infants are born with the knowledge that certain things—a soft touch on the cheek that cues the search for the breast, the spacing of two eyes on a contrasting field that centers the universe—are worth attending to. A newborn's attention starts almost at once to be reshaped by the actual stimuli presented by the environment, becoming an unfolding capacity to search and to learn.

There is a simultaneous development of the capacity to ignore and discard that continues through childhood. You can sometimes watch a mother shopping or chatting in some public place, absentmindedly emitting a stream of unenforced injunctions and teaching her children not to listen to her: "Bobby, be quiet," "Bobby, don't touch that." Bobby quickly learns to distinguish the tone of an instruction that is likely to be enforced and ignores the others as background noise. That learned skill in not listening is likely to carry over to his response to all the women in his life—or her life, for women too may learn not to accept instructions—or instruction—from women. Inattention is as much a learned skill as attention.

I probably belong to the last generation that learned to watch television like a play in a theater, totally captivated, unaware of neighbors, chandeliers, or the framing of the stage, looking intently through the wrong end of a telescope and believing in the lives and passions of tiny, distant figures. At the end of a Christmas Day in the early fifties, spent with my father and his wife, one movie after the other, I found myself stiff and battered, overloaded with conflicting emotions. Gradually it dawned on me that no one else was listening as I was, that there was a style of attention to TV that differed from the attention appropriate to a darkened theater. Sitting cross-legged on the floor, with my back to the room, I had failed to notice that behind me on the sofa there was a steady murmur of conversation and comment, drinks and nibbles, comings and goings, an easygoing social flow of which the televised dramas were only one element.

I suspect that children today learn to maintain a secret inattention during their hours watching television, just as many adults learn to layer their attention in order to get their daily activities done. Infants seem to be riveted by television, but for older children inattention, or diluted attention, is a survival skill, and most adults seem able to conduct a desultory conversation with the TV yammering in the background. Shows are designed to compete for listeners, struggling for the preemptive stimulus, and years have to be spent learning not to suspend belief, not to focus too tightly. It may be that the greatest cost of television for children is not in the content—a mixture of information and fantasy, of the useful and the distorted—but in the ways they learn not to learn, not to attend. But perhaps this is adaptive in the modern workplace.

Some of the gender differences in the ways males and females are socialized to compose their attention may have a biological basis, but the flexibility of human potentials is such that there is no reason not to assume that all of us could include the full range in our repertoires. Yet "paying" attention in the classroom does seem to come more easily to females than to males, with the result that boys often suffer from punitive efforts to exact attention—several times as many boys as girls are diagnosed with attention deficit disorder. Perhaps the cultural emphasis on docility in females, on wanting to please, is already part of a pattern of multiple attention, in which girls include both the lesson being taught, whether it interests them or not, the teacher's response to their behavior, and the view from the classroom window. It is perhaps not so much that girls are less easily distracted as that they are less likely to be fatally distracted, to switch their whole attention away from the lesson to something else.

Increasingly in contemporary society, men and women do the same tasks, but there are still visible differences in the way they do them. Some tasks that traditionally were done by women were elaborated and turned into full-time professions by men. It used to be said that women cook and men become chefs. Women care for the sick and men become surgeons. Women sew and men become fashion designers. Today, of course, women are becoming chefs and surgeons and designers, and are able to compose their lives so that at appropriate times they can focus fully on these vocations instead of weaving them in with other activities. Still, one of the things that has been pointed out about the entry of women into the corporate world is that they often attend to process simultaneously with task—to how things are done as well as whether the goal is achieved. They notice whose feelings are being hurt and who is hesitating to voice an idea, even while working on improving the bottom line for the next quarterly report. A few corporations are beginning to value this skill, but all too often it is unrewarded.

On kibbutzim, women are almost always assigned to "women's work": the kitchen, the laundry, or the children's house, where child care is shared. This is ironic because what was characteristic of traditional women's work was not so much that women did these particular tasks but that they did them simultaneously, in a single braid. Working all day in a community laundry is far more arduous than doing the wash in the context of homemaking, emptying pockets of eloquent trash and musing about children's activities reflected in stains or fraying at the knees, gradual growth, and changing waistlines.

Concentration is too precious to belittle. I know that if I look very narrowly and hard at anything I am likely to see something new—like the life between the grass stems that only becomes visible after moments of staring. Softening that concentration is also important—I've heard that the best way to catch the movement of falling stars is at the edge of vision. Yet we only hear half the story. The command that echoes from years in school, from exhortations and from sermons is always to pay attention, not to be distracted, not to stray. Eyes on the prize, eyes on the ball, eyes on the bottom line. That same resistance to distraction, shutting out the view from the window, has too often meant pursuing some goal at the expense of the environment, of the social system, or riding roughshod over other human beings.

At one time, I was much enamored of a line of Kierkegaard, perhaps because it suggested a way of being that seemed beyond my reach: "Purity is to will one thing." Over the years, I've come to realize that those who will one thing are the most dangerous people around, even if that one thing is apparently something good. Uncomfortable as it often is to divide attention and balance competing goals, trying to put these together into some larger and more inclusive composition, that is the only way to live responsibly in the world. Blindness is likely to affect anyone who pursues a single goal, whether the quest is for the Holy Grail or return on investment, yet blindness of that kind is sometimes regarded as a virtue.

The style of attention that either ignores an issue or becomes obsessive about it leads to single-issue politics and is often focused too narrowly to solve complex problems, like trying to prevent teenage killings with metal detectors in the school yard instead of job opportunities and social justice. When Californians began to be serious about the water supply, I was struck by the way some people attacked their water consumption as if they were mobilizing for a full-scale war to be fought and finished. Alas, what was needed was not a temporary obsession but men and women who would give, consistently, over time, 5.0 percent of their attention—or maybe 0.5 percent—to conserving water. It is not enough to ignore the issue and then focus in on it or to develop specialists at the cost of apathy in everyone else.

In order to switch attention to an issue that has been out of focus, it is well to have held a place for that issue in a systemic mental model. In the same way, a cook keeps some pots "on the back burner" in the multiple pattern of attention required to produce a complex meal, and a gardener has an image of the changing impressions to be produced through the seasons, anticipating them with a multitude of tasks. Some issues come in and out of focus easily, depending on need. Some problems are ignored because they have never been defined as problems. When one is taken unawares, it is hard to tell whether attention was too narrow or the mental model incomplete.

Ecological problems often develop in areas where no one is keeping watch, no one's purposes are served by vigilance. Human communities always have some awareness that water supplies may be more or less clean, but since it has been possible to take the air for granted through most of history, air pollution took time to notice and attend to. Disease has always been seen as a problem to combat, but learning to focus on health is more difficult. In the same way, our preoccupation with deferring death has blocked attention to the learning involved in dying.

It has often been said that human beings will only start attending to the worsening environmental situation as the result of disasters, but such shocks would have to be renewed repeatedly and on a steadily increasing scale. Fear is a poor teacher. Waiting for crises, mobilizing in a major way to deal with them, then shifting attention to something else because we can only attend to one thing at a time is not the way to solve the problems of the environment. We can only deal with these things if we have multiple patterns of attention. One necessary type of attention will indeed come from the professionals who concentrate, like chefs and surgeons, on one small part of the issue, one threatened species or region, and become crusaders, but crusading cannot do the broader task. What is needed is more like the kind of attention needed for housekeeping than the total mobilization needed to put out fires. You cannot put off grocery shopping until the children are starving, and you cannot provide for them by making an emergency effort and then turning away to attend to something else. In the same way, controlling the AIDS epidemic depends on persuading ordinary people, male

and female, to give just a fraction of their attention to preventing infection, even when they are preoccupied with sex
or looking for a fix.

Life is complicated. It is simplifying but dangerous to
have one overriding concern that makes others unimportant—rage or passion or the kind of religious exultation that
seeks or inflicts martyrdom. The most striking cause of nar-
rowed attention at the national level is warfare. In a complex
world of conflicting priorities, going to war can be a tremendous relief. In peacetime, government has to balance off
guns and butter, but when a nation goes to war, it goes to
war to win, and everything else becomes unimportant. All of
a sudden, the president no longer has to be concerned with
the future of industry or education or human rights. Everyone can focus in on the supreme importance of victory. We
have rarely had a government that did not emphasize guns,
even in peacetime, which should be butter time; but in
wartime, the primary emphasis is on guns. Warfare comes as
a great relief to those who prefer thinking about one thing at
a time. It is no coincidence that the language of warfare is so
often used to focus on any urgent issue—poverty, or drugs,
or the AIDS epidemic—yet the metaphor is ill chosen, for
many of these wars cannot be won, any more than a homemaker can definitively win a war against mess. The world
doesn't stop while a war is taking place, either, and the victories won on the battlefield leave other problems unsolved.

Even in warfare, there are issues of attention. Intelligence depends on the skillful use of peripheral vision. Strat-
egy depends on recognizing change. Battles are often lost by
attending too much to the lessons of previous conflicts, too
little to the present. The United States went to Vietnam with
fixed ideas of how to fight and was blindsided by the effec-
tiveness of guerrilla warfare. From that debacle it drew the
conclusion that wars should be fought only with narrowly
drawn and highly specific goals, but such goals have the
effect of narrowing attention in an unpredictable world.
Focusing in 1991 on getting Iraq out of Kuwait, the generals
ignored issues that we will be dealing with for years to come.
The war not only left Saddam Hussein in power in Baghdad
but also created a major environmental disaster in the gulf

and a major human disaster for the Kurds, and it gave politi-cal legitimacy to Syria's Hafiz al-Assad, for in warfare it is easy to ignore the bad habits of allies. Wars almost always have unintended side effects, and goals may need change along the way. Winning is never as simple as it seems.

In warfare, domestic issues are left untended. Men have on the whole had the privilege of walking out the door and assuming that they could delegate many of life's concerns in order to concentrate elsewhere. Whether for the period of a workday or a military campaign, someone else would take care of the children, the laundry, the elderly, tonight's din-ner, calling the plumber, getting on with the neighbors. Today women are meeting the demands of outside jobs, and some of these other concerns are beginning to be shared. It was interesting, in the criticism of the Bush administration that followed Operation Desert Storm, to notice that the electorate was increasingly feeling that it is unacceptable to ignore domestic issues during wartime, for often there are parallels between the household division of labor and the national priorities. This new insistence on the domestic may represent more than the familiar postwar spasm of isola-tionism.

Building peace, like women's work, is never done. A woman's work is never done because, although a particular task may be completed, she is always engaged in multiple tasks, long- and short-term, cycled and recycled, and there is never a moment when she can say that no task is waiting. The first proverb arguing against focused attention that comes to my mind is from the kitchen: A watched pot never boils. The reality is that the pot will boil, but the tea tray won't be ready. A homemaker cannot keep up with the full range of tasks by focusing on one thing at a time. In the same way, the health of a nation is always many stranded.

I think, for instance, of a village woman in Iran, in a household with several children, probably at least one other woman, and one or two elderly people. In the course of a day's work, she prepares food; she keeps track of the children and the old people, including looking after them if they are sick; and if she has an hour here or there she sits down at the carpet loom, which is right in the center of the household.

She can be interrupted at any moment, yet after a few months she has a valuable carpet, which is a unified work of art. At the end of a given day an onlooker might say that she had achieved little, because everything she did was enfolded in these interlocking patterns. Even more easily, the onlooker might dismiss the human and social value of the constant flow of conversation and gossip through the day that keeps the system functioning smoothly, but this too is essential to her living.

Women don't stop caring for children when they start cooking dinner. The current ideal of "quality time" for women who follow a schedule of work outside the home suggests that mixing tasks is inferior even within the home, yet doing a task like shelling peas or raking the lawn alongside a child often makes for a deeper companionship than stripping the moment down to a single focus on relationship. By contrast, many men left in charge of children convert that into a full-time activity. Even though bored, they reject the suggestion that they could have cleaned the house or prepared dinner in the same time period.

We live in a society that follows the industrial model in dividing activities up rather sharply, assuming that people do one thing at a time and portion out their attention in the same way, perhaps because models of value and achievement are based on activities that have been largely reserved for males. Yet the reductionism involved impoverishes everyone. If only for tax purposes, we are forced to label activities as work, or play, or learning, or therapy, or exercise, or stress reduction, missing the seriousness of play, the delight of good work, the healing that happens in the classroom. For adults, learning is rarely the only activity going on, but it is always potential. By emphasizing a single thread of activity, we devalue the learning running throughout.

I believe it is important to provide a vocabulary that allows men and women whose lives do not follow the compartmentalized model of a successful career that our society has developed to value their achievements. Life is not made up of separate pieces. A composer creates pattern across time with ongoing themes and variations, different movements all integrated into the whole, while a visual artist combines and balances elements that may seem disparate. When I called an earlier book *Composing a Life,* I imagined the cover as a classic still life that would defy the concept of separate spheres: maybe a mandolin, and some apples, and drafting tools, arranged apparently casually but actually very artfully on a table. That is the kind of combining and arranging we do in our lives. The mandolin and the fruit may come from different aspects of life, but the art is in the composition that brings them together.

The way the Iranian village woman structures her attention to combine her different activities and occupations is complex but not original with her. She watched her mother doing it, just as the way farmers and shopkeepers dealt in the past with their multiple tasks was also learned from available models. But the way we now reassemble parts that have long been separated has to be invented as we go along, with an extra layer of creativity and an extra layer of learning, putting together the human rhythms of rest and effort, combining the satisfactions of a promotion or a bonus with those that the Iranian woman gets from serving a meal or completing a carpet.

To attend means to be present, sometimes with companionship, sometimes with patience. It means to take care of. Its least common meaning is to give heed to, for this meaning has been preempted by the familiar *pay attention,* which turns a gift into an economic transaction. Yet surely there is a powerful link between presence and care. The willingness to do what needs to be done is rooted in attention to what is. The best care, whether by a parent or a physician or a teacher, is founded in observation or even contemplation. I believe that if we can learn a deeper noticing of the world around us, this will be the basis of effective concern.

In school, even the word *attention* is corrupted. If I use it in the title of a lecture, audiences don't come—particularly on college campuses. Attention is transformed from openness and delight into life's first exaction and tainted with boredom. Yet to compose lives of grace we need to learn an artful and aesthetic pattern of attention to the environment of those lives, attention that turns and turns again, embracing nature in all its diversity and other persons with all their potentials. We need a broader vision, to match the world in which we act with an image that includes the forest and the trees, the baby and the bathwater.

8

Longitudinal Epiphanies

I LEARNED ABOUT BOREDOM in the college chapel of the Ateneo de Manila. Not that I had never been bored before, but I had never before understood that boredom is learned and that other kinds of attention might be learned instead that would reshape life profoundly.

For months I had been dropping in at the college noonday masses, after teaching an eleven o'clock class. One of the Jesuits would come and hurry through the service in a big, bleak room, temporarily fitted out as a chapel. This was in 1968, when liturgical renewal was barely beginning and most of the liturgy was still in Latin. The priests mumbled, and the responses were hardly spoken by the congregation, four or five students and some older women from the neighborhood, busy with their rosaries, perhaps a dozen people in all, scattered far from the altar. At the same time, although I felt very much an outsider, I found myself going back repeatedly, watching and wondering and musing. Not bored.

With the beginning of summer, one of the younger Filipino Jesuits took over the service. He created a stir by consecrating in English, which was still not officially allowed in the diocese of Manila, and by giving the communion cup to the laity. He got students to bring guitars, and soon the chapel was crowded at noontime, full of enthusiastic, folksinging students, gathered in a circle around the altar.

Almost every day he introduced some novelty, interspersing comments or slight alterations in wording. Then one day in the crowded chapel, with sun streaming through the side windows, I noticed absentmindedly that nothing particularly different was happening: the same old phrases, the same old simple chords. Suddenly slack and indifferent, I thought, But this is dull, why on earth go through this same boring thing day after day . . .

Boredom is so familiar that we rarely recognize that we are trained in it, addicted to a consumerism of the spirit, jaded to need ever more vivid diversions. Activities we once did not expect to find full of novelty and stimulation get recast as entertainment—and then become burdensome when entertainment flags. Some peoples eat the same foods at meal after meal, but we have learned to expect not only varied meals but varied mustards and vinegars. Getting up in the morning, taking a shower, brushing one's teeth, eating a bowl of cereal, hundreds of such peaceful activities have been tarted up with flavorings and music and gadgetry, so that after a brief period of novelty they become not bland and comfortingly familiar but irritatingly boring. Boredom does no doubt occur naturally, not only for humans but for other mammals, under circumstances of extreme lack of stimulation, but we have made every area of life subject to an acquired pathology of attention.

There have been a series of experiments in educational television, devoted to packaging reading or geography in the frenetic cadences of quiz shows, cartoons, and commercials. It is hard to criticize programming that seems to work in conveying something useful, but children who are given chocolate milk to get calcium into them grow up as chocolate eaters, not as milk drinkers. Children prepared for school by children's television arrive better prepared for the content of their lessons but perhaps less tolerant of the rhythms of reflection and multiple return appropriate to gradual growth in understanding, for attention that is exacted tips over easily into boredom, while learning flourishes on the subtleties of recycled attention. Recognizing that education should be enjoyable rather than punitive, we sometimes attempt to alleviate boredom by making bits and pieces of education entertaining, instead of discovering and supporting those modes of activity to which the experience of boredom is simply irrelevant. When a people at war become mirror images of the enemy, the war is already lost.

In Manila, I was able to watch a microcosm of a struggle throughout the Christian world to reshape forms of worship; the movement for liturgical renewal prefigured some of the efforts in education today, and suggests some of the dangers to be avoided. It was fashionable for a time for couples to seek greater authenticity by composing their own wedding services, but words that cannot stand repetition year after year provide brittle shelter for the shifting understandings of marriage. It was easy to dismiss as reactionary diehards those who wanted to stay with the old forms, without analyzing what was being lost, but loss there was. During the same period when the mainline denominations were seeking greater simplicity, intelligibility, and participation, young people responded to these losses by learning to chant in Sanskrit or to daven in the style of the shtetl. I remember a conversation during that period with an elderly nun who spoke of how very difficult it was to get beyond the words of the new English translations so she could really pray.

It is not obvious, from the outside, what form or language is most likely to support prayer, or what relevance the question has for secular people with no interest in praying. The behaviors we call religious cover a curious spectrum, from the ecstatic to the formal: we see, on the one hand, Hasidim dancing with the Torah, God-intoxicated Sufis, Pentecostalists crying out in strange syllables and convinced that these are words of praise . . . and, on the other hand, rigid repetition of memorized forms and gestures. Many rituals seem moribund, but the death of ritual might also be a death of delight—or rather the loss of form and courtesy as entryways to learning and participation. If this is true, the problem affects us all.

A longitudinal epiphany seems like an oxymoron, for we are losing the capacity for epiphanies played out through time, like those that allow a man and a woman to enjoy having breakfast together day after day for forty years or to enjoy the leaves falling exactly as they did last year and the year before. When that capacity is lost, is it lost forever? Do the books and magazine articles that instruct women on how to keep husbands and lovers from becoming bored with them ever help, or is the battle against boredom lost as soon as it is joined? The struggle against boredom is, surely, like an arms race—one of those processes where the attempt at correction increases the problem.

The damage is most obvious in the arts. The search for originality dulls the capacity to savor small variations. The flight from boredom condemns both artist and audience to increasingly mediocre performances, for good work takes patience and polishing. During the Beat era and the decade that followed, when would-be artists rushed to fuzzy and undisciplined work, it was fashionable to associate art with chaotic and ungovernable novelty rather than care and discipline. Today this fashion is fading. Most of the artists at the MacDowell Colony get up in the morning and go to their studios as doggedly as business commuters, following a planned path or systematically working their way into a pattern of variations. Sometimes the Muses do visit. Sometimes the divine descends to bless the most routine prayers. But as Gerard Manley Hopkins said of the peace that so often eluded him, ". . . when Peace here does house / He comes with work to do. . . ."

Rituals work in different ways: "a sacrament effects what it signifies." Sometimes they transform children into adults, single people into couples, commoners into kings—transitions so profound they must be faced in ignorance, phrased as mysteries. More often, ritual is a sort of metaphysical housework, intended to sustain some continuity in the world. Many peoples perform their rituals and ceremonies again and again in the conviction that the sun rises, the tides ebb, seasons come and go, and the game continues plentiful because they are doing their part, in dance and song and prayer, to sustain these rhythms. A rain dance is not so much an instrumental way of causing the clouds to open as the human part in the orderly pattern that includes the coming of rain in its season. When the ceremonies lapse the natural order sickens.

Rituals use repetition to create the experience of walking

the same path again and again with the possibility of discovering new meaning that would otherwise be invisible. The tales that are told again and again of how the world began, how the Pilgrims landed, how Mom and Dad met and fell in love, grow in meaning rather than become boring simply because they have been heard before. Something that is said to have happened once becomes multiple by being evoked repeatedly in different contexts as ritual and myth give birth to each other. Many of the stories told in this volume are stories of discovery, of hit-and-run epiphanies, but the retelling offers a different kind of learning experience. Stories of unique and startling events—visions and miracles—are common in mythology, but to recount them is to claim a share in continuing truths. Many good rituals shaped by time have sections into which the new or the personal, the informal or even the ecstatic can be inserted, so that what is ungoverned and spontaneous fits into a repeated form that feeds upon it.

The timeless must flow into time to suggest a way of being in the world. This is expressed in the curious and paradoxical word *practice*: saying the words whose meaning you do not yet fully know but whose form may someday be so habitual that it feels spontaneous. It is not only in religion that learning can link the rare moments of sudden understanding with gradual change through practice, the longitudinal epiphanies. To practice—anything, playing the violin, extracting DNA in the laboratory, rock climbing, doing Zen meditation, bathing the baby—is to repeat what appears to be the same action over and over, attentively, mindfully, in a way that makes possible a gradual—almost imperceptible at times—process of change. Practice links the greatest virtuoso with the child beginning piano lessons and refutes the notion of learning as a single, one-way transfer of useful knowledge, a replacement of the unknown with the known. Communities of practice blur the line between aspirants and adepts because both are still developing. To attend—even when attending means sitting on the sidelines, like a medical student watching a surgical procedure—is to become a participant.

The tasks that have provided the basic textures of human life, like farming and child care, can be experienced as menial and repetitive, but for some they are really gradual paths of learning, forms of practice that deepen from day to day with the piquancy of minute difference. From the outside, it seems that actors must find repeated performances deadly—the same lines and gestures, night after night—but they say that every performance can be different. The rejection of repetition puts the possibility of practice at risk.

The liturgical movement was much concerned with authenticity or sincerity. When an action is sincere, we seem to believe, it comes spontaneously from within. Following a formula, as in reciting a prayer from a book or repeating prescribed gestures, would seem to be the antithesis of sincerity, yet such repetition can be the beginning of practice, so that words from a book eventually come from the heart. In much apparently spontaneous prayer, old words with the power to link the user to the faith of past centuries are paraphrased—only to become hackneyed as soon as they are spoken. Worry about sincerity is almost as self-defeating as worry about boredom, creating an internal division and spilling out to corrupt the arts and the texture of human relationships. Love must be expressed more often than it is felt. Living in unfamiliar cultures, learning to feel and express culturally appropriate emotions, I have been nagged by the issue of sincerity, yet this is a singularly American concern, which only arises in the context of a belief in some autonomous inner self separate from interactions with others.

Western expatriates in Iran often criticized elaborate Iranian courtesies and the emphasis on keeping up appearances and the dangers of losing face as insincere. Yet these coexist in Iranian culture with a deep concern about integrity that is related to but not the same as American notions of sincerity. For one thing, honor and shame and public respect (*ab-e ru*, "water of the face," which can be taken away or spilled) are not entirely individual, for a man's vulnerability to disgrace is connected with his whole family and especially his female relatives. Iranians regard disparities between inner and outer as an expected result of social life, yet they express a persistent nostalgia for a sort of purity (*safa-ye baten*) in which these are in harmony and a persistent suspicion that others may be dissembling. "The whole of ignoble conduct lies in falsehood," said Kay-Ka'us ibn Iskandar, an eleventh-century Persian writer. "The essence of truth is the negation of ambivalence."

Weeping is a good example of the relationship between feeling and performance in Iran, for weeping is taken as evidence of overriding emotion breaking down the disparity between internal and external, yet it is also a major ritual activity. One of the principal religious activities of Shiite Islam, the form commonest in Iran, is mourning the deaths of the imams, Muhammad's descendants through his daughter Fatima, believed by Shiites to have inherited the leadership of the community and persecuted through the centuries by the majority Sunnis. Weeping for the imams gives religious merit and earns their intercession on the Day of Judgment, so on the anniversaries of their deaths men's and women's meetings (*rowzeh*) are held in mosques and homes as the stories are retold, interspersed with prayers and singing, offering the chance to have "a good cry" and feel refreshed. All who are present sob or moan and beat their breasts, and tears flow freely. Men, for whom these mourning activities provide a community network throughout the year, march in procession or organize dramatic reenactments of the battle scenes leading up to tragedy, the women gasping and pulling their children back as horses gallop by.

When I went with women friends to such mourning gatherings, it was clear that they were expecting to weep and half-expecting me to weep as well, for they seemed to feel that the deaths of the imams, especially the martyrdoms of Ashura, were so tragic that weeping would be a natural response, even for a nonbeliever. But there were plenty of signs that repeated weeping is learned and managed, with a skillful leader acting rather as a cheerleader. In wealthy houses, ornate Kleenex holders spaced around the room are part of the furnishings—silver set with turquoise, I saw in one house, where afterwards a maid came around with an ewer of rosewater. Between sessions, cheerful gossip prevails, in no way subdued, but when each session begins, weeping takes only seconds to get going. Once I saw a toddler worriedly trying to lift his mother's veil as she hunched over on the floor, sobbing, her veil pulled entirely over her, but in general children watch solemnly without weeping and go home to playful imitation, with their mothers' good-humored approval.

It is unfair to call this ritual weeping insincere, for it absorbs and transmutes the genuine griefs of life. At a mosque the day before Ashura, I heard a woman break into convulsive sobbing, so stark that the leader stopped her narrative and the other women looked up and fell silent. My companion whispered to me that the woman's son had been killed in a political protest a few years back. After a few minutes of dithering and consultation, the women drifted out. The ceremony was resumed some time later, but for the moment it was shattered, whether because the bereaved mother's emotions were ungovernable or because they were politically risky, I could not tell.

In a sense, all mourning is mourning for the imams. The preaching at burials and memorial services is full of references to the martyrs of Shiite Islam, echoed in the ebb and flow of audible weeping and the expectation that one will be moved to tears even at the funeral of a relative stranger. Poor elderly widows take their griefs to funerals often: not exactly professional mourners, they may be given gifts of cash. The bereaved are supposed to be overwhelmed by grief—they throw dirt and ashes on their heads and wail. Men as well as women who suppress their grief are respected the less for it.

Between the death and the burial, however, those who are close to the deceased or to the family go, dressed completely in black, to the house. Their behavior could not be more different from the gregariousness of a Filipino wake. Entering through an unlocked door, friends sit in silence, separate and self-contained, eyes cast down, with only the barest acknowledgment of greeting. Tea and cigarettes are offered, but the theme is isolation, a turning away from social form and ceremony, an ascetic of grief. Here familiar rituals serve in a different way, for against the background of flowery Iranian greetings and elaborate hospitality, this austere reversal of ordinary social behavior is extraordinarily moving. You have to have forms to dramatize their inadequacy.

Weeping in public outside ritual contexts, like overwhelming rage, is both feared and admired in Iran. It is scandalous to be so overcome by emotion that one is unable to maintain appearances but almost admirable to have emotions strong enough to escape the dividedness of self-consciousness and calculation. In popular films, heroes may be so carried away by rage that they ignore all risks in the effort to achieve revenge, or so overcome by love that they will even marry a woman who is not a virgin(!).

Tears flow and flow again in courses etched by culture. Ritual weeping in Iran is apparently "Method" weeping—it depends on a learned skill in drawing on real emotion, not on the ability to simulate tears. Christians have regarded tears shed in memory of the crucifixion as a divine gift, and Jews gather at the Western Wall, all that remains of the temple in Jerusalem, to mourn its destruction. The cadences of poetry or phrases of music in a minor key can bring tears to the eyes; eyes may sting and a lump form in the throat at the wedding of a stranger. At age three, Vanni turned to me at a poignant moment in a Disney film and said, "Mommy, why do my eyes have tears?" We seek these moments out and are refreshed by them, small and recurrent instances of catharsis.

Ritual mourning is a form of practice in both senses, for all peoples learn to use social forms both to mask emotion and to focus it—as ways of going about feeling the right feelings and performing the right actions, ways of achieving that intensity of response which is itself unifying. During the Iranian revolution, individuals risked their lives to walk ceremoniously in processions for the dead that were affirmations of political protest.

Other people's rituals may be bizarre and inaccessible to empathy, because they have been shaped and elaborated over time, encrusted with unintelligible detail become sacred. One of my mother's students in a class on field methods, who had grown up in a family where there was no religious observance whatever, was told to attend a religious ceremony of some group not her own and take notes. She took herself that Sunday to a Greek Orthodox mass, but some twenty minutes into the service, through which she had dutifully been writing down everything she saw, she was struck by the extraordinary and to her horrifying thought "They do this every week." After that she sat, bemused, taking no more notes. Someone with no previous experience of ritual can learn to grow through the recurrence of the incompletely understood, but the necessary patience is not often taught. In order for ritual to lead to learning, ritual itself must be learnt.

Some rituals are explicit about the learning process, like the Passover Seder, in which the ritual includes the asking of questions by the youngest person present. The first Seder I attended, in 1957, when I was seventeen and living in Israel, was in the home of a Yemenite rabbi. A scholar of Arabic at the Hebrew University had arranged for the invitation, explaining that hospitality to strangers is appropriate at Seders, which echo other festive meals in the Middle East. He pointed out the piles of lettuce leaves to be dipped in sweetness and the baked lamb eaten as a main course, contrasting them with the more schematic elements of an Ashkenazi Seder, the exotic compared with the ordinary, although both were new to me. The form followed is based on the Haggada, a text composed of multiple layers, some in Hebrew, some in Aramaic, reflecting recurrent experiences of conquest and liberation. The layering process continues; today the story is retold in dance and drama, in nonsexist or even feminist versions, with the horrors of the holocaust and the dramas of Zionism woven into the layers of historical memory, and the events of the twentieth century prefigured in the exodus.

The annual celebration of Passover has fixed elements and a fixed sequence, reflected in the schematic design of the Seder plate on my desk, but its meaning is subtly altered each year by the new historic context, just as historic events gave new meaning to the Shiite martyrdoms. The Seder is also different for every person at the table, for at any annual ritual the experiences of the year past are spiraled into dialogue with hopes and memories, generating some new measure of insight as life cycles through the seasonal round. Weddings and funerals recur as well through a lifetime of coming to terms with endings and transitions, and even weekly and daily rituals contain the possibility of growth as they return over the same ground in new ways.

In American culture today, we acknowledge the usefulness of some kinds of repetition and the comfort of habit, but still we regard repetition as tedious and artificial, scanting both rituals of courtesy and those greater and ancient repetitions that affirm and sustain faith, preferring words never before spoken that come spontaneously to the lips. The formality and repetition of previously elaborated and sanctioned materials are contrasted with unique moments of visionary or ecstatic experience. Yet ritual, like my daily walk through the woods in New Hampshire, can be like Heraclitus's river, never the same. Repetitive ritual and ecstatic expression, like continuity and discontinuity, are two sides of the same coin rather than opposites. Religious forms cluster around extremes of decorum and anarchy, formality and spontaneity. These seeming opposites often coexist in the same tradition, or grow one out of the other, as in Quakerism. Both make it possible to bypass or get beyond what can be spelled out linearly in reasoned prose. The familiar word *enthusiasm* once referred to such outbreaks of undomesticated religious fervor.

My best opportunity to observe this complementarity came in the early seventies, when I joined a group of about twenty young, white Protestants who met weekly to pray in a suburban living room near Boston. This was what is called a "spirit-filled" or charismatic group, and I was drawn there by a double curiosity. I had been wondering, as a linguist, about the phenomenon called speaking in tongues, or glossolalia, which has been interpreted since the very earliest days of Christianity as a gift of the Holy Spirit. I was curious too about the movement of a style of worship associated with poor black and Appalachian churches into the middle-class mainstream, for beginning in the sixties small groups like this one began to appear in a range of Protestant denominations and in the Catholic church as well.

Speaking in tongues turns up in the New Testament description of Pentecost (which is related to the Jewish feast of weeks, Shevuot, as Easter is related to Passover). The disciples, we are told, "were all with one accord in one place. And suddenly there . . . appeared unto them cloven tongues like as of fire, and it sat upon each of them. And they were all filled with the Holy Ghost, and began to speak with other tongues, as the Spirit gave them utterance" (Acts 2:1–4). The passage goes on to describe how Jews visiting Jerusalem from every country of the Diaspora heard the Galilean followers of Jesus preaching to them in their own languages. Peter then stood up and preached to that first great revival meeting, sometimes regarded as the founding of the Christian movement. The understanding at that time was that glossolalia involved believers speaking intelligibly in languages they did not know.

Glossolalia seems to have kept turning up as the Christian movement spread, sometimes becoming a source of disorder and contention. Paul's often quoted paean to love (charity) is part of a scolding he is giving to the church in Corinth for being more interested in dramatic manifestations than good human relations: "Though I speak with the tongues of men and of angels, and have not charity, I am become as sounding brass, or a tinkling cymbal" (1 Cor. 13:1). We memorized that passage at my school, but we understood that Paul was saying that love, which he goes on to describe in words of great beauty, is more important than eloquence, not that he was warning against the disunity created by spiritual fireworks. By that time, it was clear that the ecstatic utterances of glossolalia might correspond to no known human language, so they were ascribed to the languages of angels.

The story highlights some of the paradoxes of religion: ecstasy and altered states of consciousness in conflict with community order. A vision that transcends differences becoming a source of conflict. The dream of standing up and communicating to people of all nations in ways that make sense to them is as different from the labored communication of the United Nations as the ancient dream of flight is from being stranded at Heathrow. It would be nice if the diversity of modern society looked more like the first Pentecost and less like the church in Corinth or the construction site at Babel.

Paul disapproved of unintelligible utterances in common worship but accepted them as a form of private prayer, of communion. Similar behaviors are not limited to Christianity: they crop up all around the world, often associated with the idea of possession, demonic or divine, and often representing a kind of proof of direct inspiration. What they surely prove is that most individuals are capable of altered states of consciousness when the culture supports them, ways to move beyond the dreary doubts and blindness of the ordinary. Sometimes they inspire strength in adversity, leadership, extraordinary self-giving; sometimes they inspire self-righteousness and fanaticism.

The evenings I spent sitting on a sofa in that suburban living room full of young charismatics were another path to wondering what happens when different codes and roles are mingled. There was considerable diversity in that group. Some were working class, others were graduate students; some were preoccupied with a vision of love and others with a vision of judgment. Some seemed to manage their shifts of consciousness well, having found a context where they were approved and valued, but there were others who, I suspected, were unstable and erratic in ordinary life.

Learning to speak in tongues, like becoming part of any community, is a powerful psychological transition, involving trust and surrender, group pressures and fear of loss of control. Becoming a participant here involved learning the rhythms of a group with no clear leader, putting together the intelligible and the unintelligible: wild emotion alternating with pauses to pass cake and iced tea and chat about the weather. I discovered that there was much that was routine in the group's spontaneity, a recurrent limited vocabulary. I wondered whether members felt estranged from themselves when they lifted up their voices in words that felt involuntary, as if coming from somewhere else. Later, I visited black Pentecostal churches as well, shy and imprisoned in my whiteness, puzzled by the simplest questions of how to participate, like what time to arrive for a nine o'clock service.

The concern for spontaneity is based on the assumption that real feeling moves outward from within, but emotions often work their way inward from actions or arise from learned forms and symbols. The slow loss of the capacity to use form with and for conviction is not only a loss for society but ultimately a loss to the individual's own capacity for experience. During one of the World War II studies of Japanese culture, an interviewer made a comment to a young Japanese woman about Japanese respect for the father. "Oh no," she said, "in Japan we do not respect the father." The interviewer was flabbergasted, all his expectations contradicted. "You see," she said, with the most delicate emphasis, "we *practice* respect for the father." "Why do you do that?" he asked. "In case," she said, "we someday find someone who deserves that respect." Often practice precedes experience and makes it possible. With the decline of those interpersonal rituals called courtesy has come not an increase in "authenticity" but a decline in individuals shaped by these forms into personifications—and models—of the virtues they have rehearsed.

Some ideas can only be passed on through participation in sound and movement, through art or ritual. There must also be ideas which can only be arrived at, only be invented, through such activity. Dancers, from Martha Graham to Zorba, have said, "If I could say it, I wouldn't have to dance it." Patterned motion on a basketball court or a dance floor invites participation in fluid form that may even serve as a medium for shared thought, inviting practice and offering the pleasures of skill. Indeed, for many contemporary people, sport and exercise, devoutly pursued, are the best available models of practice. Abandoning ritual we become like children who go through an important early stage of life in a language which is then forgotten, so that a body of experience and thought remains frozen and inaccessible to review. Some kinds of understanding grow only in repeated participation in forms that are not fully understood. Answers are given at the Seder to the child's question, but the Seder teaches more than these answers.

I learned another lesson about boredom in the Philippines. Doing fieldwork in the Marikina Valley, I brought with me the disabilities of American society. I was coming from a culture in which every moment is likely to be flavored with extra stimuli, a magazine to read on the bus, music in the elevator, television or radio dissipating the musing peace of early morning. Too much stimulation had jaded my palate, putting every activity under threat of boredom. But in the village my commitment to learning gave me an alternative stimulation, for every moment became interesting. All the trivial chitchat, repeated over and over, whereby people reaffirm their connection to each other, was part of my growing understanding of the culture I was there to study, a ritual I gladly learned to join.

It was not until after I ended my research and left the field that I understood my disability and how the role of anthropologist had helped me compensate for it. I used to go back and visit, dropping in without advance notice, to reaffirm my appreciation of the kindness of those who had put up with this awkward stranger, rather than simply take my notebooks and disappear. But now I went with a new kind of purpose: I went in order to have gone, to have checked in.

There would be news of things that had happened in the village, and sometimes I had news to tell about myself, but I remember those visits as deadly boring. My pleasure at the texture of day-to-day village life had depended on learning, learning that required sustained attention and a degree of continuity, not on savoring the ungarnished present. One of the joys of language study is that everyone and everything encountered in the new language is exciting, and this is true of other patterns of behavior that an outsider needs to unravel. When I let myself abandon the effort of learning—finding the patterns, putting the small details in context—I became easily bored, and at the same time I felt inauthentic, cut off from these lives I had wanted to know.

Perhaps in old age with a rocking chair I will find myself contented in the moment, able simply to drift and enjoy the ebb and flow of social life, as the villagers seemed able to do. For now, the nearest I come to the quiet listening of others is lulling a baby, shelling peas, sitting and looking at the flames of a fire or the flow of water in the stream beneath my window. I could pursue that quietness through the disciplines of meditation, whether as a religious practice or for personal growth, and hope it would overflow into my styles of participation. But looking back on those attacks of boredom and discomfort, it is clear to me that I could also have gone on cultivating the habit of learning and taking pleasure in noticing whether people stood or sat, how they moved their hands, the small, perennial variations in talk about the weather. I have survived many committee meetings with the same doubled attention, so that boredom was simply beside the point. I am convinced that, regardless of theology and dogma, this too is a form of spirituality.

9

Turning into a Toad

I GREW UP LEARNING NATURAL HISTORY from my father, paddling in summer through the swamp in New Hampshire or wandering the California woods, absorbing his steady attention to plants and animals. In the city, I kept an aquarium and went with him to the zoo. But when we moved to Iran, doing natural history with Vanni came less easily. The mountainous desert around Tehran has, of course, its own flora and fauna, but it was hard to get to with a small child and unfamiliar to me, dry and thorny. The zoo was a scandal: tatty, miserable animals, mocked by visitors offering lighted cigarettes to the chimps and reinforcing in their children the sense that wild animals and even most pets are vicious and dirty. Persian culture values cultivated places, walled and irrigated gardens, so when Iranian families picnic farther afield they bring their rugs, the images of enclosed and hospitable spaces, along with pots of food and samovars.

Vanni and I found our wildlife where we could: tadpoles in garden pools, beetles, a hawk with a broken wing that lived in our greenhouse but was never able to fly again. Vanni was out shopping with the housekeeper one day when they saw a large tortoise, its shell cracked in some highway encounter, and Vanni picked it up and put it in the shopping basket. The housekeeper was half hysterical when they got home, afraid to remove her passenger or to abandon the groceries. Vanni and I set the tortoise loose in our little garden, where its shell healed in the course of a year and it grazed peacefully on pansies and the excess kitchen greens the housekeeper began to enjoy hoarding for it.

Some months later, I traveled to Europe for a conference, my first absence since the family had come to Iran. I got up to wait for a cab to the airport while the rest of the household was still asleep. The early morning is a gentle time, before the dew is dried and the sun blazing, so I stepped out in the garden to greet the tortoise (whose name was Mud) and the day, and there I saw a magnificent toad, a toad that I very much wanted Vanni to see. I put it in a large empty storage jar and set it by her bed, rushed to scribble a note to my husband to make sure it was released later in the day, and ran out to meet my taxi.

And I began to worry. What would a two-year-old, waking up to find her mother gone and a toad sitting beside her bed, surely conclude? What price Piaget? What price Kafka.

A phone call from the airport woke Barkev up to move the toad away from the bedside so it could be produced later in the day, and confusion was averted. But I have wondered since whether I should have left well enough alone. Perhaps the transmutation of mother into toad might have conveyed a significant truth, a lesson rather than a trauma. Toads are quite close to human beings, after all. Not so close as the primates, of course, or even as a sheep. Not what you might call first or second cousins, but perhaps what large southern families call kissing cousins. The frog prince is much more plausible than man into cockroach.

In 1990, a colleague at George Mason University, Harold Morowitz, was involved in a first major conference to consider the significance of crashing populations of amphibians—frogs, toads, salamanders—around the world. Observations had been trickling in from herpetologists (amphibians are studied by the same experts who study reptiles), each one noting a drop in the population of a particular species in a particular ecosystem: fewer tree frogs in the Costa Rican cloud forest, a muting of peepers in Connecticut suburbs, a dearth of toads in the Himalayas. It was no easy matter to connect these observations as part of a single pattern, for herpetologists, like anthropologists, are territorial creatures, and each one had a specific hypothesis to explain the trend in his marsh or her stream. They are no doubt still debating the details, but the trend seems to make sense only in global terms, for it has been observed far from construction or factories or toxic dumps. The most likely cause is the increased ultraviolet radiation in sunlight, but the increased acidity of the rain or the dissemination of highly stable and lethal manufactured molecules could also have worldwide effects. Perhaps all of these have combined.

Frogs and toads, because they live in both air and water, and because of the relative permeability of their skin through which, in addition to gills or lungs, they take in oxygen, may be especially sensitive to environmental change. Because of their similarity to humankind, they have been compared to canaries in a mine shaft, a kind of analog computer registering the implications of changes for other vertebrates, but at a higher degree of sensitivity. We can look at amphibians and wonder about ourselves.

There is another kind of similarity between amphibians and human beings, a long period of barely protected development. There are many solutions in different species to the problem of how the zygote formed when sperm and ovum meet is to be protected through the stages of development until it can survive on its own. In mammals, the crucial steps take place inside the mother's body, followed by a period of adult care; baby birds go through their early development protected by hard shells, and butterflies go through their complex metamorphoses protected by cocoons. But the amphibians start out encased only in a moist, soft jelly, then go unprotected through the dramatic changes familiar in the transition from polliwog to frog, growing limbs, absorbing tails. These are all changes that humans go through in the shelter of the womb. You can visualize the vulnerability involved when you recall the impact of thalidomide when it was introduced into the protected internal environment.

Among the mammals, human beings emerge from the shelter of the maternal body least able to function independently and require the longest period of adult care, not only developing physically but also learning the distinctively human patterns of survival. It has been argued that one of the trends in human evolution has been a reduction in post-partum physical development, by a sort of lowering of the physical standards of maturity: less body hair, less heavy brow ridges, slighter secondary differences between males and females. This trend is referred to as neoteny and may be an essential part of the continuing flexibility that underlies adult learning, for we are not only naked apes but, through-out our lives, infantile apes, learning and sensitive. The wrench of letting children go to determine their own lives is a recognition that they are not yet mature and perhaps never will be.

The analogy of toad to human includes the recognition that both are mutable, both realize their distinctive charac-teristics in intimate contact with their environments. Vanni would have had a good deal of reason on her side if she had looked at the toad in its roomy jar and said, That could be my mom; the herpetologists who see the crash of amphibian populations as a warning to human beings have multiple lay-ers of truth on their side. But many of them read the evi-dence more narrowly, and most of us decline to see its rele-vance. Having learned to think of child development as a process of differentiation, I thought that the association of mother and toad was to be avoided. I lost, perhaps, the chance to establish a pattern of metaphorical thinking, espe-cially the kind of metaphor called empathy, with the deepen-ing of attention it offers.

Vanni might have learned a number of other important lessons if she had kept the toad in a jar by her bedside. For one thing, she would sooner or later have had to deal with its death. Most people encounter death first in a pet or a road kill or a dead bird found by the door and only gradually dis-cover that death is a common heritage. It is not easy for chil-dren to learn that parents are vulnerable, fallible, mortal, to learn the necessity of protecting even while being protected. It is important to learn that when you think you own a frog or a houseplant or even a biological community like a forest or a lake, although you can destroy it, you may not be able to keep it alive. Human conventions of ownership permit the possibility of destruction, but they do not guarantee the ability to preserve.

For a long time it has been assumed that the natural world is knowable and that the capacity to know equals the right to use and rule. Knowledge is a form of domination and a step on the way to more practical kinds of domination, but in practice domination has preceded understanding. We work out our relationships with other species across a gulf of incomprehension, even of those that have been studied for centuries and are maintained under the most artificial conditions. When many species meet in a complex ecosystem, their balance is often beautiful to the observing eye, but the orchestration is elusive, and the fit that makes possible a joint performance bridges differences far greater than the mutual intelligibility between members of different cultures.

We began from the image of a Persian garden as a setting in which plants and animals have been cultivated and controlled, a setting in which people of different backgrounds and understandings encountered one another and managed, for a time, to join in a single performance. The walls had been built to exclude wildness of all sorts: wild animals, desert weeds and weather, nomadic raiders, the predations of rulers and neighbors. But there is still a wildness inside the garden. Even in Persian carpets, modeled on gardens and like them the work of human hands, there is a hint of the unknowable in the tradition that every carpet should contain a deliberate error, for perfection belongs only to God. Through the centuries of cultivation, gardeners have learned to manage mysterious processes they could not understand or see. Some of these have one by one become transparent to scientific instruments, but then new knowledge reveals further layers of mystery in their interactions.

An ecosystem, even that largest of all ecosystems, the biosphere of this planet, is not created by knowing the properties of all the parts and fitting them together, nor do we as human beings know all the steps of the complex dance in which we participate. This was demonstrated by a large and curious experiment in Arizona called Biosphere 2 (Biosphere 1 being the one we all inhabit). Biosphere 2 is a kind of Persian garden. It is a huge, sealed dome erected near Tucson, within which a dynamic model of the planetary biosphere was constructed, containing a carefully designed diversity of organisms of different kingdoms in appropriate ratios—plants, a few animals, a small group of human beings—in the hope that they would establish and maintain the cycles needed to survive over several years. But the most important lesson of Biosphere 2 may be the unpredictability of the process, with surprising fluctuations at every level, from the mix of gases in the air to the hearts of the human inhabitants.

The planet we inhabit, larger and more complex than any model, has its own ancient stability, yet it continues to change, new species evolving and the old shifting into new constellations. Even as we try to describe the properties of the system, new ones emerge—and so do new theories of the nature of unpredictability. Often enough the pattern is changed by human actions, but the results are not fully controlled by us. Natural systems, encountering change, often meet it in unexpected ways, seizing unvisualized opportunities. Every antibiotic-resistant strain of bacteria, every rapidly varying virus, from flu to HIV, is affected by its environment, both in the short term and as lineages—and the immune systems of organisms, in their varying complexity, learn at the same time. The AIDS epidemic is not a demonstration that the natural world is inimical or judgmental but a demonstration of its essential wildness as the changing habits and interactions of human beings destroy some species and create new niches for others. When we think we understand what we are seeing, knowledge itself proves to have unexpected properties. Strangers meeting in a walled garden, whatever mental models we have of the place where we meet, these must include an acknowledgment of the unknown, a quality of tentativeness that allows us to feel our way into relationship. Yet without models, without metaphors and analogies—poetry and mathematics and experimental constructions like Biosphere 2—it is almost impossible to think about the planetary ecosystem as a whole, and thus impossible to care.

The parable of the blind men and the elephant is endlessly repeated to illustrate the problem of different and partial points of view. Everyone knows it: one blind man is standing by the elephant's leg. He feels it up and down and says, This is a creature rather like a pillar. Another one is off at the front, and he finds the trunk and says, No, it is rather like a snake. One finds the tail and reports that an elephant is rather like a rope. Each one has some sort of detailed knowledge based on contact with a part of the elephant, but they have no way of putting it together.

From time to time I have played with the fantasy that if I were in that situation, with a group of blind men or women, each of whom had knowledge of one part of an elephant, I'd go out and find a small mammal, a puppy perhaps, and put it in turn on each one's lap. I'd be able to say, That pillar you felt corresponds to this part, but now you can feel, because this creature is smaller, the relationship of legs and tail and nose and eyes and so on. All terrestrial mammals are constructed on the same ground plan, so there are very great similarities between a dog and an elephant, sufficient to use familiarity with the dog as a framework for putting together the specific knowledge that each person has of a part of the elephant to make a whole. There is a pattern that connects the dog and the elephant, largely the same pattern that connects the mother and the toad, the child and the sacrificed sheep.

However many times we hear about that elephant with blind men groping from one end to the other, the elephant never moves. It is always, in the stories, an immobile elephant. A puppy, on the other hand, is likely to wiggle. At the same time that the ground plan of a mammal would be conveyed, giving a context for inserting specific specialized knowledge, a much more immediate sense of the aliveness of the creature would come across.

A puppy would be a living metaphor. It is important to avoid the rhetoric of merely and not disparage the very complex truth that it would convey, for metaphors are what thought is all about. We use metaphors, consciously or unconsciously, all the time, so it is a matter of mental hygiene to take responsibility for these metaphors, to look at them carefully, to see how meanings slide from one to the other. Any metaphor is double-sided, offering both new insight and new confusion, but metaphors are not avoidable. By recognizing similarities, metaphors bring different kinds of knowledge together, but they also preserve the pockets of mystery that are part of the whole. The puppy is not fully known; it brings its ambiguities to the understanding of the elephant. Even the special strains of laboratory animals used to model human frailty are still largely mysterious.

In the contemporary effort to attend to the relationship of our species to the biosphere, we are combining multiple metaphors, old and new. The dips and crashes of amphibian populations may look like a rather minor side effect of modern developments, peripheral to the human dramas we tend to focus on, but a recognition of similarity between our kind and theirs brings these remote dramas into the same story. On a much larger scale, the Gaia hypothesis, developed by James Lovelock and Lynn Margulis, who adopted for the theory the name of the Greek goddess of the earth, asserts that this planet is alive. This integrates a vast amount of information in a single image: What we are talking about is life. It wiggles. It may bite. The metaphor provides a bridge from high technical specificity to all the experiences that go with direct contact with a living being.

The Gaia hypothesis is not a simple assertion that could easily be proved or disproved. Instead it is a complex statement with multiple levels of meaning, like a work of art. At the simplest level, it asserts that this planet is characterized by the capacity for self-correction that characterizes living organisms, the maintenance of continuity by corrective variation. The composition of the atmosphere and the temperatures of the planet over time, for instance, are not accounted for by the laws of chemistry and physics alone, but apparently by processes of self-regulation within those laws. The Gaia hypothesis deepens the sense of the planet as developing, as having a history. Wiggling.

There are several quite different responses to the idea of the earth as a living organism. Some people feel increased solicitude for a planet newly recognized as vulnerable, intricately beautiful. Others respond with nonchalance, saying Gaia can do her own housekeeping, leaving us free to continue as we are. Still others notice that the planet's capacity for self-correction might well involve the end of the creatures that are making the trouble.

When a metaphor is proposed it generates questions. You notice the puppy's ears, and you wonder if the elephant has ears. You didn't notice so much hair on the elephant—something else to wonder about. A metaphor goes on generating ideas and questions, so that a metaphorical approach to the world is endlessly fertile and involves constant learning. A good metaphor continues to instruct.

When we assert that the planet is living, one of the ideas that springs to mind is that living things can die. They have needs that must be met. Their health is subject to thresholds of various sorts. We may extend the analogy and move quickly to wondering what part we represent in this organism. Are we, perhaps, the brain? That feels good. Are we the vectors of its reproduction, colonizing outer space? Or perhaps its immune system? Perhaps we are a virus running wild within it or the multiplying cells of a malignant tumor. As with any pathogen, the question arises whether this one will kill the organism it invades, or be eliminated or neutralized, or whether some balance will be achieved. Microbes ride the human body and depend on its health as we ride the earth. All these speculations and others come up in playing with the metaphor of earth as a living organism. A metaphor can propose testable questions and serve as a framework for synthesizing information. This is no small gain, for the challenge to meteorologists and chemists, physicists and geologists of synthesizing their knowledge far outdoes the problem of the blind men in the old story.

There is another step implied in the Gaia hypothesis, the hint that the behavior and characteristics of this planet are best grasped by an analogy with the living organism we know best, a human being. Perhaps after we've handed the puppy to the blind man and the puppy has wriggled free and escaped, we will persuade our friend to feel his own nose and legs and notice that he has four limbs. He can use his own body as the model.

The Greeks believed that every tree was inhabited by a dryad, a female spirit, who would die if the tree were cut down. Most of us know intellectually that trees are alive—

they grow, they age, they breathe, they respond to the environment they're in, they draw in nutrients, they have a metabolism, all those things—but it's hard not to slip into seeing a tree as an inanimate object that is simply a given in a particular environment. Trees live at a different tempo from human beings—I didn't slow down enough to notice the growth of trees until I was over forty—so it is hard to remember that a tree can suffer and become ill, though we are more aware than we used to be. Arguably, a belief in dryads may complement what is learned in botany classes, making our knowledge of trees more complete and more accurate. In the same way, the belief that patients are whole persons is not easily acquired in medical classes that emphasize the mechanical characteristics of bodies, so physicians must find other ways to maintain it. Although the belief in an immortal soul brings a lot of baggage that may be troublesome, it probably helps some physicians to remember. The anthropomorphic dimension of the Gaia hypothesis proposes empathy as a way of knowing—and caring.

Above all, the Gaia hypothesis evokes the powerful ancient metaphor of Mother Earth. In the early seventies, there was a poster of the earth as seen from space, the picture that has become so familiar and beloved, and underneath was written, "Your Mother—Love Her or Leave Her!" That was a brilliant but confusing poster, because every young American male knows what he is supposed to do with his mother: Grow up and leave her. After all, his entire socialization is geared to achieving independence. The poster fed right into the fantasy that if we messed up this planet we could climb into spaceships and zoom to another one or perhaps to a space platform. No one had done the arithmetic on how many people were going to fit into the spaceships. That poster was an invitation to believe in the possibility of leaving, in the self as separate and separable.

Since the early years of the space program, the fantasy of solving environmental problems by leaving this planet behind has faded, as has the metaphor of earth as a spaceship, but we still may not have found the metaphor that leads to effective attention. A metaphor can obscure as well as reveal. In contemporary American culture, I doubt that the best way to elicit caring and responsible behavior from adults is to remind them of childhood, the retrospective dumping ground of problems and resentments. I may feel that having the earth thought of as female enhances me or allows me to empathize a little more deeply, but I hate to expose the planet further to the danger of rape or evoke the ambivalence that people feel about mothers.

The use of a personal name, Gaia, suggests that the planet can evoke the attitudes we reserve for identified human individuals. Do we love Gaia? Does she love or trust in return? What does it add to understanding or confusion that Gaia is the name of a deity from an ancient and polytheistic system no longer widely worshiped, the most primitive layer of Greek religion? The original Gaia was inclined to devour her own offspring, many of whom were monsters.

Perhaps we could empathize more constructively if the metaphor were differently conceived. Because the life span of a planet is potentially so long, we might learn to think of the planet as a young child that requires care and attention but has an unknown future. Such a metaphor would underline the need to protect future possibilities, not only for our human descendants but for all life on earth, and might make accepting the limitations on knowledge and control less painful.

When we use a metaphor that is drawn from human relations, it is well to look carefully for all its hidden implications, for we run the risk of evoking human conflicts. If we are going to think of the earth as female, it behooves us to take a good look at gender relations, because gender relations of dominance and exploitation will infect, have already infected the relationship with the planet. Images of children often do evoke protectiveness and caring, yet we have been willing to incur massive debts our children will have to pay and all too many parents exploit or abuse children and even more feel they have a right to determine a child's future. If we are going to use family images, let us take some responsibility for constructing human families that offer metaphors of mutuality and hope.

To me, the most important thing that the Gaia hypothesis proposes that was absent from earlier metaphors like spaceship earth is that we are immersed in, brought into being by, a living reality, not a mechanical one. We are completely dependent, as we would be in a spaceship, but we do not have full blueprints and we cannot expect to be in complete control. The atmosphere, that mixture of gases we study in high school chemistry, could occur only as the product of a living system, for the free oxygen that makes animal life possible would not continue except for the steady activity of green plants. The soil and most of the rocks we think of as lifeless are the product of life processes over vast stretches of geological time. Because the earth is different from us and mysterious, changing constantly, every encounter with the environment is an opportunity for learning. Our planet is not an inert piece of real estate subject to rezoning, for its surface has been shaped by life processes, with their own lawfulness. We cannot treat the earth as inert, just as we cannot treat a tree as an iron pylon or a meadow as a piece of wall-to-wall carpeting.

Environmentalism began with piecemeal concerns about parts of the natural world: saving this forest or that bay, the whales or a particular lake that had become polluted. The issue becomes very different when we realize, as the Gaia hypothesis demands, that we are totally contained in and sustained by a single living system, in which all the parts are interconnected and everything we do resonates with the whole. Nothing is fully localized. The destruction of an ecosystem or a species is an amputation and, like the amputation of a limb, can trigger fatal shock or, at the least, require learning new ways to function. One extraneous item introduced in the wrong place in a living body can trigger pathology. The Gaia hypothesis becomes, at every level of its metaphorical evocation, a reminder that the world we live in is a biological, or if you like a biologized world, a sacred process in which we share, a community to participate in, not an object to be used.

We don't see it. Our habits of attention work against seeing, and the connections in the system are invisible. Most of the time, we are like the blind men with the elephant. Focusing on the pursuit of particular, narrow goals, we pay attention to a fraction of the whole, block out peripheral vision, and act without looking at the larger picture. Cutting down forests for timber, it is easy to ignore their role in the regulation of climate. Poisoning insects to increase crop yields, it is easy to ignore the concomitant deaths of natural predators, which lead to an increase in pests the following year. All of this could be spelled out in environmental impact statements with elegant diagrams showing the interconnection of the different factors, but the Gaia hypothesis may help to make these interconnections seem intuitively obvious. Environmental politics has looked very like other kinds of politics, many groups dealing separately with different obsessions rather than with a single, interdependent whole. Certainly practitioners have to concentrate and certainly there have been victories, but it seems unlikely that lasting stability will be won with piecemeal approaches.

There is a curious kind of blindness in the way we formulate our conscious purposes and press ahead, blind to consequences, blind even to those consequences that eventually loop back and affect the overall viability of the system. My father, Gregory Bateson, wondered whether this phenomenon was a flaw in the perceptual system of our species or only a culturally induced blindness produced by religious traditions or the mechanistic view of the natural world developed in the Enlightenment. He was especially suspicious of the Cartesian partition between body, that which we study and manipulate, and mind, which he associated with patterns of organization and self-regulation but which Descartes saw as transcendent.

The Enlightenment represented a moment in the way philosophers describe the world, but human beings have always been blind in many ways, for like every other species we are characterized by the sensory limitations that match our evolutionary adaptation, fitting us to a hunting-gathering lifestyle. We are blind and deaf and otherwise insensitive to a great deal that other animals can perceive. There are patterns on the petals of flowers like landing pads for bees to lead them directly to where the nectar is, but because they are outside our visible spectrum we cannot see them without extra instrumentation; the sensory capacity to receive and process that information is not part of our biological equipment. The puppy we have imagined sitting on the blind man's lap can smell and hear many things that the man cannot. These capacities have to do with how the ancestors of dog and man and bee made their livings. The toad has the appropriate perceptual system for catching flies.

In fact, human beings have a very broad sensory range and a learned capacity for narrowing in. We learn patterns of attention, ways of concentrating, and things to watch for, often culturally defined. We can also modify these patterns of attention, as the blind man may develop extra acuity in other senses to compensate for the loss of sight, and we can learn to see through multiple lenses. How might human beings become sensitive to the effect of their actions on the toads and salamanders? The very ease with which a child can imagine kinship with other species is a starting place for understanding as valuable as the disciplines of the laboratory.

The Gaia hypothesis pulls the data together, but it goes further by offering a metaphor for organizing awareness of the interconnections. Beyond that, it proposes empathy as a way of knowing and imagining connections about which we cannot yet be explicit. It cannot, however, guarantee love or respect any more than centuries of religion and philosophy have been able to end the exploitation by human beings of one another. We continue to be unable to provide adequate care either for the old, our parents, or for the young, our children, to whom we will entrust the future, so it is no wonder we mistake the planet that represents both source and destiny for a shopping mall. What would it be like to walk through the woods or the city in the presence of—aware of—Gaia? Part of that awareness can be built up by letting children look through microscopes, germinate seeds, learn about soil chemistry, but part of it comes into being through the experiences of loving and being loved, resolving quarrels, learning new ways of family life, attending patiently to things we do not understand.

What does it take to notice when the peepers or the bullfrogs aren't as loud this year as they were a year ago, as Rachel Carson did with birdsong? We have a name that we use for noticing without being sure of the exact cues, the details that might be offered as evidence. We tend to call that intuition. Most of what is called intuition has to do with information processed subliminally, unconsciously. "Mere" intuition it is not very respectable, like "mere" metaphor. Yet even though the boundaries of analysis will be pushed ever further, it is a mistake to discard the hints and suspicions that are not accounted for by a given paradigm.

All thought relies on metaphor, on ways of noticing similarity so that what has been learned in one situation can be transferred to another. Scientists try to purge metaphor and intuition from their publications, but freed from the formal constraints of scientific publishing, the speech of scientists is like all human speech and thought, full of metaphors, often unconscious and unexamined.

The solution is not to purge metaphors from speech and try to ignore them; the solution is to take responsibility for the choice of metaphors, to savor them and ponder their suggestions, above all to live with many and take no one metaphor as absolute. There are truths to be discovered in equating one's mother with a toad; there are truths to be discovered in looking at a butchered sheep and recognizing heart and lungs and death itself as common. We have work to do to make empathy an acceptable form of learning and knowing for people who are not poets and therapists. We have to make it possible for manufacturers and politicians to admit empathy as a legitimate, conscious discipline, thoughtful empathy as a form of knowing, leading to effective action.

It would be well to have our vision of the planet based on metaphors that would evoke no blindnesses, no impulse of exploitation, but there is too much blindness and exploitation in ordinary life for such clarity. We will not arrive at the point of treating the planet with respect until we are able to treat all the members of our own species with respect. In the same way, we will not be able to treat the natural world with respect as long as we lie to ourselves about ourselves, so it will be important to become accustomed to the reality of death.

Human beings are economical in their patterns of thought, transferring ideas from one context to another, using a multitude of unconscious metaphors. Not long ago, in 1988, a group of parents in Tennessee brought a lawsuit protesting that their children were being taught the religion of "secular humanism" in the schools and objecting to the use of fantasy and mythology in education. A picture from a reading primer that showed a little boy and a little girl sitting at a kitchen table, with the little boy putting a piece of bread into a toaster, was cited as undermining traditional concepts of the family. This may seem extreme, yet these parents were right in their understanding of how people think and learn. Not only does such a picture undermine traditional concepts of the family but it undermines traditional concepts of God, for male dominance over females has long provided a model for the relationship between God and humankind. They would also be right to resist the metaphor of the dryad, along with any other suggestion of sacred presence immanent in the natural world, as undermining the idea of God as transcendent, ruling from outside and above.

Family systems, the organization of institutions, the way we run our country, the way we respond to other cultures and races, and the uses of political and military power—all these things are based on interlocking sets of metaphors. Our many relationships are isomorphic: they have the same form. There is a pattern that connects, and it is a pattern of dominance and exploitation, taught again and again in the most ordinary human arrangements. That pattern is expressed in the fierce and ultimately self-destructive attack on this planet that we cannot rule because we are a part of it.

Developing visions that protect the earth is going to take long-term work, not just on what we do with sewage, or how we deal with smoke coming out of factories, but on how we handle our interpersonal relations. In effect, because knowledge and perception are so dependent on available models, they cannot be changed without a commitment to changing basic patterns of social life. This is the most significant sense in which we are our own metaphor.

How ironic it is that, in the story of the Garden of Eden that is so often seen as having crystallized for millennia the notion of human dominance over the natural world, the Fall is a punishment for eating the fruit of the tree of knowledge.

Yet even after they have eaten that fruit, it is clear that Adam and Eve do not understand God or apple trees or serpents or even each other. Expelled from the garden, they set forth on the path of misunderstanding—and denying—their own desires.

10

Joining In

IN THE PHILIPPINES THEY DO A DANCE called the *tinikling*. Two people kneel holding opposite ends of thick bamboo poles, using them to beat out a rhythm, two beats apart, one beat together, while dancers run in, stepping between the poles on the beats apart, nimbly leaping out when the bamboos come crashing in toward their ankles. Filipinos are skilled at making sure that foreigners try their hand—or their shins—just as they are skilled at making sure, preferably before a large, convivial audience, that foreigners eat *balut*, fertilized duck eggs incubated to the point at which the embryo ducks, still soft boned, have recognizable bills and feathers. In this society where shaming is a pervasive and effective sanction and the statement "I was ashamed" an often given reason for action or inaction, considerable graceful skill is available to chivy or wheedle overbearing foreigners into feeling obligated to make fools of themselves. When you watch Filipinos dancing the *tinikling*, what you see is skill, one of those dances in which virtuosity is central. Under the pervasive charm, the "smooth interpersonal relations," as social scientists translate the delicate Pilipino word *pakikisama*, there is plenty of malice to enjoy pratfalls or occasional hits of the loosely held bamboos.

When Barkev and I got married in 1960, one set of dancing styles held sway in the United States, with variations depending on age and class. By the end of our years in the Philippines and in Iran, dancing in the United States had changed radically, ceasing to be an activity in which a male led and a female followed, one which a well-bred young lady could neither initiate nor decline, and became something that men and women, boys and girls, did in their own individual styles, with very loose patterns of coordination, sometimes alone, sometimes with partners of either sex. "Knowing how to dance" still included the possibilities of virtuosity, aficionados going to clubs night after night, but gone were the anguished failures of stepping on a partner's toe or losing one's place in a complex jitterbug step. Many of my generation never made the transition to dancing to rock music—generations cling to the melodies and dance styles of their youth, and for us learning to waltz or to tango once led into pleasure and elegance. It is easier to replicate the product of learning than to find the continuities in the process.

At the MacDowell Colony, an evening of listening to music by a resident composer and looking at slides of the work of a painter was followed by dancing. Joining in, I found myself caught up in an enactment of what I had been writing on the nature of participation where codes are not fully shared. Dancing is increasingly a matter of improvisation on the dance floor, some limiting themselves to a narrow range of motion and still rather rigid, others finding new flexibilities and rhythms. For generations used to treasuring ballroom dancing as a rare opportunity for physical contact, it seems isolating, even autistic, yet there are plenty of ways in which dancers can turn their separate improvisations into dialogues, not so much alternating as simultaneous, like conversations in sign language. Rather private styles can be linked by changing orientation in complementary ways, choosing to mirror the movements of the other, or alternating rhythms, just as it is possible to establish a game with an infant by sharing gaze, echoing vocalizations, passing a toy back and forth.

Changes in dance styles reflect other changes in society. We have come a long way since the minuet, constructing forms that allow everyone who will to participate without knowing a precise, shared set of rules, without rehearsing the details of fancy footwork. We have moved away from the notion that someone (generally male) must always be in control of any shared activity. Even in marriage we are developing different ways of harmonizing styles, with variable degrees of coordination. Not surprisingly, when the wider change is contested, one generation may condemn the dance styles of the next, labeling them obscene or chaotic. Yet the dance style that is evolving offers a new kind of commonality to a society of increasing diversity and ambiguity. Styles of dance, like the forms of ritual, convey very complex notions of how a society is organized. The new styles, never quite spelled out, are learned through participation and carry the lessons of a new style of learning that is crossing and recrossing the oceans and spreading through the life cycle.

The problems of an era are likely to be reflected in its dance as well. The metaphor of compliance to a traditional form has been loosened, but now the need to belong is expressed by metaphors of increased intensity and intoxication by movement and rhythm that permit young people to drift into decisions without acknowledging that they are doing so. There continues to be conformity, tightest during the adolescent years, conformity in styles of clothing and language and imitation of behavior, pressure toward dangerous experiments with sex or drugs or alcohol. There are casualties in any process of change, elements that suggest risk and others that suggest adaptation. The youth of today, like previous generations, will dance off the dance floor and into the myriad settings of life, but they take with them a different set of models for participation.

When I went back to Israel with Vanni in 1988, after thirty years away, the first experience that drove home for me the degree and kind of change in the interval had to do with dancing. For years I had been telling Vanni stories about the time I had spent there at the end of high school, visiting on a newly founded kibbutz near the border of the Gaza Strip. When we went back together to visit, I had predicted dancing and singing on Friday night, men and women alike in freshly laundered trousers and clean shirts, many of them embroidered (we packed carefully to be ready), singing the folk songs of a newly constituted "folk" and dancing the hora, with its evocation of community and vitality, its lack of sexual reference. But the kibbutz has changed. Women have drifted partway back into traditional gender roles and styles, privacy has increased, and now the only dancing on Friday nights was an occasional disco organized by the young people, with music and dancing not unlike the rock one would find in New York.

All around the world, human beings make music and dance, using forms and rhythms that resonate with the other activities of the day. The anthropologist Alan Lomax has written about the congruence between styles of work and styles of music and dance, postures and rhythms and whether people perform separately or in unison. There is probably a relationship too between performance styles and learning styles, a match in which one supports the other. What I learned in afternoon classes in ballroom dancing was all too congruent with the assumptions of the time about propriety elsewhere. In times of change such congruence is replaced by a disruptive mismatch, in which song and tapping feet suggest a way of joining in that is contradicted in more formal settings. The successfully learned skills of the dance floor or the street corner can either undermine or enrich the more slowly changing classroom.

Dance styles reflect the everyday experience of encounters with others in which we do not "have all the moves," often because newcomers have no way of discovering a stable set of rules or because the rules have changed overnight. Arriving as a stranger at one of New York's airports, the passenger from London or Taiwan or Houston does not know the New York regulations for tolls and taxis, the names of bridges, or what the ride is likely to cost; the driver, newly arrived from Tehran or Manila, may barely know the geography; the two may have no language in common. Not all rides are completed without friction, but an amazing number are, some of them with laughter. In between, of course, some passengers are cheated and some cabdrivers are mugged.

In Tehran I spent weeks unraveling the rules for fares and pickups in the multiple-passenger taxi system that was a major source of irritation for foreigners, but I still found myself running into new rules applying to such issues as large bundles or traveling companions. In the end I discovered a solution for situations of ambiguity that let me gracefully pay the small extra amount that was possibly illegitimate without brooding on it all day. If the driver asked for an additional payment that went beyond the rules as I knew them, I would pay it, saying to him, *"Halalet basheh,"* "Let it be permitted to you," a formula meaning that whatever had been taken was religiously permitted and would not be counted as theft. It meant I would not stand up and denounce the offender on the day of judgment. It was the perfect solution: if the driver was indeed trying to cheat me, he would laugh and I would have gotten a bit of my own back; if he was innocent, he would protest. Either way, I would know the rule for next time; either way, I went on without feeling ripped off. I didn't have all the moves, but I had one that would do.

One of the oddities of American society today is that standards and regulations have changed and multiplied to the point where no one is in compliance with all of them. Even those who have never paid a household worker without withholding taxes or been present where marijuana was being smoked have driven a little over the fifty-five-mile-per-hour limit. Even those who are careful to express emerging forms of respect may be haunted by words or behavior from the already obsolete understandings of a decade ago. Of course faculty date their students, an administrator at Amherst said, how else can they find someone to marry? Even those who try to follow all the tax regulations hesitate to try to learn them, so they become dependent on some accountant's asking the relevant questions.

This is not a matter of accepting the wish to comply for the deed and doing better next time. No one says, *halalet basheh.* As a result, we have a society in which virtually everyone is vulnerable to the tyrannies of selective enforcement, and an IRS audit or a passing patrol car makes the most conscientious citizen nervous. Day after day, we are forced to play without knowing the rules in situations where small mistakes cannot be laughed off or used for further learning. One effect is to discourage participation. Another is to undermine integrity, which cannot easily flow from intention to performance. We play Wonderland croquet, and the Queen of Hearts is waiting to say, "Off with her head." This particular kind of ambiguity may or may not trigger the development of schizophrenia in the organically vulnerable, but it clearly inhibits learning.

The new styles of dancing were developed at the peripheries of society, among those for whom inclusion is problematic—African Americans, the young, and the poor. Reflecting our common dilemmas, they go on to offer a model that includes beginners without shaming them but still allows for developing expertise. It would be a mistake to develop a society in which participation was so simple that increasing skill (and imagination) carried no rewards, but it is also unwise to arrange things to create chronic insecurity. Whatever sense of self young people bring from home and school is tempered in the improvisational give-and-take these dances reflect.

For a long time, the assumption has been that the skills and rules of the game of life can and should be learned before the beginning of play. Generally, the rules are learned where they cannot be used: in the classroom, where the rewards of success are unrelated to the real application of a given skill. Some societies rely far more heavily than ours on learning through participation, as in the institutionalized apprenticeships described by Jean Lave and Etienne Wenger, or the simple permission to be present and occasionally join in, but all societies rely on something like apprenticeship for very early learning. You learn to speak your native language by listening subliminally to thousands of conversations and by joining in where the adults are fluent and you are permitted imitation and experimentation, gaining the rewards of effective communication and not being excluded for mistakes. In language classes in school, students do not usually gain real rewards for getting their message across—food when they are hungry, rapport, a budding romance—instead, they get A's and B's for detailed correctness.

Schools do not exist in never-never land, though it sometimes feels as if they do. At the same time that children may or may not be learning through formal instruction skills to be used in other settings, they are always learning through

participation: how to be students, how to survive in the peculiar school environment, how to be a child of a particular
race and sex in that era. They are learning how to learn but
at the same time how not to learn, what not to learn, and
who not to learn from. Teachers, for instance, may be
marked off as people not to learn from, just as working-class
servants in upper-class British households must somehow be
marked off as people not to emulate, in spite of constant contact. Learning to spend the day not learning is no great feat
when we realize how effectively most cultures convey to
young males that they should not be like the persons who
most often care for them. If you, as a young aristocrat, speak
like the gardener or nursemaid, you will not be a lady or gentleman. If you, as a male child, go too far in imitating your
mother, who is likely to be the person with whom you spend
the most time and who does the most informal teaching, you
will be a sissy. In many settings, if you as a black or Native
American child learn from your white teacher, you are a
traitor. Children who fail in school are often the ones who
have learned the implicit lessons of participation all too well;
the A students may be the failures. Furthermore, knowledge
and skills may be acquired never to be used except on exams.
Every individual probably learns some behaviors and skills
that it is not appropriate to display. Eskimo women know
how to build snow houses, but don't.

Most people are unaware of the intricate structure of
what they have learned from participation, of the intellectual
complexity of common sense or the unstated pattern of courtesies that make Emily Post and Amy Vanderbilt sound like
primers. The assumptions of everyday life are highly
abstract; spelled out, they sound like a philosophy lesson.
Conscientization and *consciousness raising* are terms used for
the empowering growth of awareness of how society really
works. They might equally be used of the intellectually
empowering process of becoming aware of the range and
depth of knowledge acquired through participation and
observation.

Turning social occasions into minefields of possible sole-
cisms is a splendid way of enforcing a class system—the

angry young man from the slums will never get the heiress because he uses the wrong fork and fails to pick up those witty references to the classics that must not be allowed to bounce before they are returned. He betrays his lack of cultural—or class—literacy in differences so pervasive that he appears to be a different order of being. When I was a child, my mother sent me to ballroom dancing classes—not because she felt that they were very important or even that the social life they were part of (cotillions, junior assemblies) was important, but because certain skills were a precondition of participation if I should ever want to do so. I learned both lessons: to dance (badly) and to regard social forms as things that could be learned when they become important for participation. Forty years after those tedious afternoons in velveteen frocks and Mary Janes, counting steps and bumping knees with little boys who loathed the whole performance even more than I did, when I told Vanni I didn't know how to dance like her friends and increasing numbers of mine, she offered me the chance to learn by participation, saying, "Sure you do. Come on, let's try it together."

Of course it is good that symphonic music is not swamped in improvisation and that we sometimes sit and watch the most rigorously trained ballet dancers. It would be well to do so more often. But some kinds of participation have become more elusive. It is not good if the availability of canned expertise on records and videos makes us all increasingly intolerant of amateur performance, of doing things less than perfectly. Many people now use their singing voices so rarely that they are shy even of singing lullabies.

You can compare religious traditions and see a direct connection between the possibility of simply joining in or its absence and the effort to include. Orthodox Judaism is a thicket of detailed injunctions, Biblical commandments elaborated during centuries of prohibited proselytizing, functioning to limit interaction with outsiders. At the opposite extreme, Islam, still the most rapidly expanding of faiths, demands little immediate knowledge from those who would convert. The convert is permitted to enter and then to learn by participation, although there are plenty of detailed regulations and abstruse theological ideas to be pursued later, and the regulations do effectively separate believers from nonbelievers.

Learning by doing is coming back, not only as an alternative educational doctrine but in areas where competence is most highly valued. Doctors in training are being brought into contact with patients earlier and earlier, so that learning becomes life-giving rather than grade-earning. Computer jocks play constantly at the very edge of their competence, seeking new paths rather than repeating correct ones. Perhaps in desperation, we are constantly inventing new kinds of instruction, and some of these are interactive and support experimentation. Computers and video games shape new forms of practice and attention.

Traditionally, human beings have approached many of the most important activities of life with very little explicit training, but when households included grandparents and uncles and aunts and multiple siblings, there was a range of idiosyncratic models available. In today's small and isolated families, new parents run the risk of bringing an infant home from the hospital with neither models nor experience to guide them. Adults may hear the news of a malignant tumor and have to improvise a response without ever having been close to another human being in the process of dying. The media air matters that used to be private—but private worlds used to be larger, so new forms of information and instruction may be the only way to benefit from the experiences of others in crises where participation is not optional.

We are engaged today in a rapidly improvised mortal dance with our planet, in which we are going to have to work out ways of protecting the natural environment without full information—without, for instance, certainty even of the degree or significance of global warming. All behavior has potential impact. It is not possible to sit on the sidelines, saying, I can't dance. Our dealings with the planet have always included actions taken before the results could be predicted, and this is unlikely to change, in spite of the achievements of science. Rather than assume everything should be mapped out beforehand, we might do better to develop rapid and fluid styles of midcourse correction. Little that we do is without risk, but not all risk is culpable. Assuming that all risks should be controlled in advance may lead us into danger rather than the reverse. Everything we do is subject to error.

Living in a society made up of different ethnic groups offers a paradigm for learning to participate without knowing all the rules and learning from that process without allowing the rough edges to create unbridgeable conflict. The clues are often elusive, like the laughter of a Tehran cabdriver, but the belief that one already knows the rules or that a god-given set of rules exists can quickly create a stalemate.

In Tehran we organized an interdisciplinary study group of Iranians and Americans. Each of us was present in multiple capacities. We were all living and working in Iran and could be self-observers as well as observers of others. We all had subjective and personal experience to offer as well as training in some field of observation and analysis. The Iranians in the group had more to offer in the way of reminiscence; the foreigners had more to offer by way of contrast, plus a steady harvest of tales of embarrassment and perplexity. The group was quite deliberately modeled on a common Iranian pattern, the *dowreh*, a kind of group that meets at regular intervals for everything from bridge to political conspiracy to reading poetry—so, if nothing else, Barkev and I were learning the grammar of a different kind of participation. Our particular group met to discuss Iranian national character and, inevitably, to enact it at the same time.

Nothing was regarded as too trivial. In a story about a man salvaging his friendship with the new husband of his ex-wife, I learned the phrase *halalet basheh,* which I came to use with cabdrivers. Conversations moved rapidly from subject to subject in a sort of shared free association, including gossip and gleeful anecdotes of childhood mischief, following long strands of metaphor. Verbally we danced together.

One day when we had gotten going on the subject of quarrels, a member of the group described how, as children, siblings would be *qahr* with each other: ostentatiously refusing to speak or even to recognize the presence of the other, until some third person interfered. That created a zestful

flurry of stories and explanations of terminology, as the conversation moved to anecdotes of *qahr* behavior between
adults, often relatives or neighbors, and to an exploration of
the role played by the outsider, someone older and trusted,
who can bring about a reconciliation (*ashti*). The conversation wandered to other examples in which physical violence
was flamboyantly threatened in some public quarrel and
bystanders could be relied upon to come between the com-
batants.

We were struck by the contrast with conflict in American
society, where often no one has an interest in mediating, so,
if the participants cannot agree, no one else will bring them
together. Even in marriage, an American husband and wife
can quarrel and split up with relatively mild effects on the
social fabric, although of course the cumulative effect of
many such conflicts is very significant. In Iran the relationship between a husband and wife is scarcely private, for it is
likely to affect the day-to-day affairs of multitudes of kin.
Handling conflict and resolution by triggering the interfer-
ence of a third party makes sense in a society where individuals are multiply connected, so that both tiffs and profound
conflicts become the business of others. Even in Iran, when
two people who are not part of a network quarrel, becoming
qahr does not lead to resolution; it simply means they no
longer speak to each other and the relationship is over.

We had a lot of conversation about *qahr* and *ashti* and
laughter about childhood quarrels, children ostentatiously
snubbing their mothers, and so on. Iranians tend to see the
refusal to interact as aggressive, but it also sends a signal to
concerned third parties that they had better interfere and
prevents the speaking of unforgivable words. Argument,
debate, face-to-face encounters will only make things worse.
Conflict resolution in Iran involves something that might not
be expected from observing Iranians in other contexts: a
reduction in expressive behavior, an asceticism of emotion
reminiscent of the reserve shown on entering a house following a death.

The traditional forms of a *dowreh* gave us a framework
that was both freeing and constricting, for it was hard to add

new people once the group settled down. We used to rotate our meetings from house to house, and I remember talking about *qahr* in a garden as we drank tea and nibbled at fruit and pastries. Afterwards, Hassan, a British-trained psychiatrist, would prepare a traditional water pipe for Barkev. Tea ended the sequence of the day. Conflict, anger, hurt feelings—these too are subject to cultural rules, even though most behavior in situations of conflict is also improvised. In America today we are going through a curious revival, in the conflicts of marriage, of the conventions of chivalry, as couples are tutored in fighting without mortal wounds.

In the autumn of 1979, not quite a year after Ayatollah Khomeini's triumphant return, the American embassy building in Tehran was taken over by a group of young Islamic zealots, and the staff was held hostage. The crisis rapidly turned into a standoff, and it was well over a year before it was resolved. During that entire period, news anchormen conducted a count-up of the days the American hostages were imprisoned and focused on every snippet of news about negotiations. In Tehran, anti-American demonstrations were choreographed for the media while spectators sipped soft drinks as if at a parade.

I had presented a scholarly paper sometime earlier on the conventions of *qahr* and *ashti,* and I drew on it during the hostage crisis for an op-ed piece in the *The New York Times* about Iranian patterns of crisis resolution. Whether the piece had any effect, I do not know, nor do I know whether others made the same points, but I learned from the problems I kept running into. At the *The Times,* they wanted to edit out my comment that a degree of press silence might be a necessary component of *qahr* behavior to make mediation possible, since Iranians would regard American press commentary as a component of official communication. The piece was titled "Doing It Iran's Way," but at the State Department I was told that there was no reason for the United States to modify its behavior since Iran was in the wrong. We approached the crisis with few alternatives, for convention made them nearly unthinkable. Yet the resolution of the crisis did come, too late to save the Carter presidency, after the

administration had suspended public comment and after a
third party, the Algerian government, had taken on the role
of mediator.

It is not surprising that a model for a possible solution to
the problem could be found in very humble kinds of domestic conflict, which diplomats were unlikely to look at. Very
humble and, in Iranian eyes, rather childish. These are the
settings in which everyone learns from experience how the
world works, acquiring assumptions so deep that they may
never be stated. Even highly sophisticated political thinkers
can be trapped by what they take for granted about how the
game is played. One of the American assumptions is that
conflict is resolved by direct negotiations and openness is
preferable to secrecy. Older and wiser, the Iranian tradition is
inclined to believe that face-to-face meetings are more likely
to cause trouble than to resolve it and that positions repeatedly proclaimed in public leave little room for maneuver. Trying to discuss these alternatives may seem to fly in the face
of common sense or decency, to be so perverse as to be
intended to fail.

The conflict between different sets of assumptions crops
up everywhere, sometimes wrapped in the elaborate protocol
of international negotiation and sometimes in a taxicab or
ten years into a marriage. We need to invent forms of interaction that will allow for learning without confessing ignorance and for mutual accommodation without either participant surrendering. It is not helpful to expect Iranians to
accept Western styles of negotiation, but it is also not helpful
to tell Americans to abandon their beliefs in order to interact
more effectively, for "open covenants openly arrived at" is
still a principle honored in America and freedom of the press
a jealously guarded right. My efforts were complicated by the
fact that although I could reach Americans to tell them how
to understand and perhaps use Iranian conventions, I had no
corresponding way to get the same kinds of messages to the
Iranians.

How does one combine in a dance or play a game when
there is no common set of rules? On the dance floor, without
learned steps, the easiest way to develop a shared pattern is

to mirror the other, and here it works. Such mirroring often occurs in conflicts however, when parties who have different backgrounds and interests start imitating each other's worst habits. Of all games, warfare is the one in which the partners are most easily tempted into symmetry. We risk becoming like our enemies as we join with them in common performances of mayhem. Revenge, tit for tat, retaliation, these are all effective and ultimately lethal ways to set up an intelligible exchange. Often in warfare one becomes more symmetrical with the enemy while increasing inequality at home. Israel is a society in which many believe passionately in democracy and in equality between the sexes, yet military preparedness supports authoritarianism, and women who are kept away from conflict (partly because of gender attitudes among Arabs, dutifully reflected) are forever at a disadvantage. Cold war concerns for security made the West mirror the East in spy games and McCarthyism.

In the Philippines, when Americans and Filipinos interacted, it was usually the Filipinos who had learned to manage English and, superficially at least, Western cultural systems, but we saw a few situations in which genuinely bicultural individuals improvised new solutions. One Sunday morning a fire broke out on the eleventh floor of a new thirteen-story office building on Ayala Boulevard, the shiniest and most pretentiously modern part of metropolitan Manila. When the fire department was summoned, however, it became clear that the pressure in the city water supply was not strong enough for hosing above the ninth floor. A large crowd gathered to watch the fire, which would have to be allowed to burn unchecked in the upper stories. The crowd included both Americans and Filipinos, many connected with the businesses in the burning building. There was no overt conflict, but there was a striking contrast between ways of responding to the situation, and a good deal of muttered commentary. I was able to watch the American response in three ways: introspectively in myself; in the Americans whose comments I could overhear; and in the behavior of an American executive whose firm rented the ground floor of the building. These responses fitted together very neatly.

In myself: I fidgeted, following a train of thought in which I fantasized appropriate action in the absence of water pressure; the only thing that occurred to me was that one might be able to enter and strip flammable curtains and carpets from the ninth and tenth floors. Listening to Americans in the crowd: they stood there and said, over and over, in criticism of the Filipinos enjoying the show, "You would think somebody would *do* something! How can they just stand there?" The American executive had taken off his jacket and tie and, without anyone helping him, was carrying boxes of company records from the ground-floor offices into the building next door.

All these exemplify the same theme, the itch to take action (and the perception of inaction as inferior). Americans tend to say to themselves (as I did), There must be something I could or should do. If they are uninvolved, they may stand by without acting themselves but assuming somebody should be acting. In this case, there really was little to do except stand and wait: my own project, as I recognized almost at once, was nonsense, since the floors involved were full of smoke and highly dangerous. The activity of the executive carrying boxes from the ground floor to the neighboring building was also nonsensical, for clearly the fire would not be allowed to burn lower than the ninth floor. Gripped by the American compulsion to take action, he looked both desperate and rather silly. He was responding, however, to a cultural need; any American directly involved at that point would have had to find some kind of action, preferably muscular exertion, whether it were helpful or even actually increased the danger. Inaction was the appropriate behavior, but it was culturally abhorrent. Americans tend to take a crisis situation as one which calls on them for a decision between positive courses of action; we seem to have difficulty even mentioning and discussing inaction as one of the alternatives. Sometimes, especially in foreign policy, Americans get into trouble by acting when they should sit still, yet in some situations this aspect of American "initiative" is a source of strength.

The impossibility of saving the upper stories was partly the consequence of projects undertaken without working out the possible contingencies, in this case the erection of high-rise office buildings where water pressure was not available for adequate firefighting, decision making not based on elaborate forecasts or alternative courses of action. This may be related to the style of Filipino "initiative," much appreciated in international working groups, a lively willingness to undertake new projects, often, however, leaving their completion rather vague. However, the appropriate behavior of simply watching was not abhorrent to most of the Filipino crowd: the fire was indeed beautiful and dramatic, and the Filipinos present could relax and enjoy it. The frustration Americans feel when they cannot take action is an internal experience that many Filipinos have not learned to have. Yet who knows but that the world we share may not sometimes be more harmed by the American tendency to overreact (or perhaps, in the Manila slang of the day, "over act") than by the Filipino tendency to underplan.

Americans and Filipinos, Iranians and Israelis, all peoples can be driven to dysfunctional behavior by their cultures; but individuals from any of these groups can learn to transcend their cultures, often because they have met others with respect and pondered their differences. After a while I noticed the executive of a company with offices in the upper floors of the burning building moving around casually in his weekend clothing. He was able to combine a graceful capacity for inaction, an exterior tranquillity that corresponded to the situation in which nothing could be done directly, with an immediate realization and pursuit of the only meaningful sort of initiative. Chatting with other businessmen watching the fire with their families, he was negotiating for rental space so that work could be resumed at nine o'clock the following morning. He was a Filipino who had had extensive contact with Americans, including study in the United States, so he had two cultural styles available and the skill to combine them in a third pattern.

Sometimes you meet people who have learned their way around the culture of the "other" well enough to have access to a second way of seeing the world. They then have a unique capacity to pick and choose among behaviors and assumptions that would otherwise have remained unquestioned, and

even to invent new ones. Tiresias the seer is said to have
acquired his vision as a compensation for his blindness, yet
in the myth he had also become wise by experiencing life
both as a man and as a woman. I have sometimes wondered
what Tiresias would have been like to dance with.

I I

Composing
Our Differences

DURING THE WINTER there is space for some two dozen artists at the MacDowell Colony, staying for periods ranging from a fortnight to two months. One encounters new arrivals in the evening, waiting in Colony Hall before dinner, reading the papers and chatting. Most colonists introduce themselves by first name only, sometimes adding an identification of their field. "Hello, you must have arrived today. I'm Joe. I'm a sculptor." "Hi. I'm Mary. I'm a poet." Some colonists have been there more than once, or know each other from elsewhere, but for most the use of first names tactfully avoids the issue of who is well-known and who is not. Everyone is working on some kind of project, so "What are you working on here?" is the next question, followed by the standard American "Where do you come from?" The questions asked are those to which everyone is supposed to have a ready answer, though I had some difficulty explaining this project or my own nomadic life. More sensitive questions of economic or marital status or sexual preference are deferred. The Colony does not provide address lists, but from time to time someone about to leave will collect addresses by circulating one of the paper place mats from the dining room, making a list to be photocopied. That's a tradition.

Max, a composer, joined the table I was at on his first day and was offered a glass of wine. "How are you doing the wine this year?" he asked. "It seems to go by a different rule every time I'm here." This time the convention was that one person or another would show up at dinner with a bottle and offer it around the table. Those who accepted would take their turn at offering another day. It is easy to imagine the pattern shifting as different groups come and go. Each new arrival is trying to decode the customs of the place, but with so much turnover it is easy to take a chance observation as a sign of ancient custom. Old-timers in my group plotted confusing rumors to pass on to new arrivals—just a trace of hazing.

Arriving at the MacDowell Colony was certainly different from arriving as an anthropologist in a Philippine village, but, because I think of artists as inclined to defy convention, I moved tentatively. This is probably as independent-minded a group of professionals as one would be likely to find, but they do echo trends in the culture of the arts. We are not in a period when being an artist implies a lot of marijuana or absinthe. In this crowd no one even openly smoked cigarettes, there was talk of aerobic exercise and of the dangers of food additives, and everyone seemed to work very hard and to be rather self-contained, cautious about emotional involvements that would consume precious time and energy. One person passed through, found himself unable to work in the long hours alone, made an effort to involve others, then departed. Watching novelist after painter after composer finding the way into this evanescent community, working out the local conventions, I was struck by how easily we met and how quickly we became attached to one another. Yet on the evenings when colonists presented their work, I was startled by the weird and wonderful ways of seeing the world hidden behind the affable faces around the dinner tables.

The easiest way to live in a strange community is to become a member of a household, often taking on a local identity by formal adoption and the granting of a name. In Israel, a society adept at absorbing new immigrants, the first question I was asked when I told people that I was going to stay to complete my senior year in high school was whether I had taken a Hebrew name, so of course I did so. Called Galila, I went to stay with a German Jewish family with three children, the oldest a girl of my age, with whom I shared a bedroom, the other two young enough not to know English, ideal teachers. Certainly I was unusual—there were not many sixteen-year-old gentile girls on their own in Jerusalem—but I had the benefit of the many conventions for dealing with new Jewish immigrants and with "the stranger within the gates." The absorption of immigrants is such a constant preoccupation in Israel that it is recognized as a major function even of military training.

In Iran, I went with Vanni to live for a while in a large traditional household. The first thing the mother did was take me to the bazaar to buy cotton print fabric for an appropriate chador, since the black veil I arrived with was too urban. She was prepared—indeed determined—to correct my etiquette and make sure that as a member of her household I behaved modestly. When I said good-bye I spoke the appropriate words and invited her and all her family to come and stay with me in Tehran, but I wondered if I could possibly welcome even one of them as they had welcomed the two of us.

Every household or society has some capacity to make a place for new arrivals, if only for the newborn. Most can absorb some other kinds of immigrants. Often new relationships are modeled on biological ones through various kinds of adoption. Sometimes anthropologists forget that they have not gotten unique access to a fixed system: the flexibility had to be there for them to take advantage of. Kinship may seem to be a fixed way of organizing society, with relationships biologically determined, but this is partly illusion—it looks that way to outsiders, especially those who have grown up in nuclear family households with only the most tenuous awareness of other kinship ties, but everywhere there is some flexibility so that outsiders can somehow be tucked in.

Among the San, because there is only a very small supply of names in use, an outsider can work with the coincidence of names to establish relationships. The anthropologist's choice of a name sets up ties. If she meets a man whose mother's name is the same as the name she has adopted, she might call him son for purposes of play or persuasion.

In the Philippines, one Spanish custom that has been taken up with enthusiasm is *compadrazgo:* at significant ritual initiations, like baptism and marriage, everyone acquires godparents who have obligations to the godchild but are also now linked to the biological parents and to each other as *compadre* and *comadre,* co-father and co-mother. A whole new and powerful kinlike network can be constructed by recruiting ritual kin for the several children of a family, useful for politics or business. And indeed it is not necessary to wait for a ritual: I have been addressed as *kumare* (comadre) by analogy, felt flattered by it, and known it would have a cost. Anthropologists who are adopted into the communities they visit find that if they try to do everything they would have done as a dutiful daughter or son there may be little freedom to circulate and little time for keeping notes. Yet even an assigned role in the system does not eliminate uncertainty about appropriate behavior—after all, those who claim these roles by right of birth vary as well. Any viable culture must deal with at least some kinds of difference, only occasionally extruding someone who cannot be absorbed. Most societies do far better than the industrial West, for instance, at finding places for those who are retarded or mentally ill.

There is a section in the Passover Haggada that explores the Biblical injunction to explain the exodus from Egypt to "thy son," pointing out that the next generation is not all the same, varying in age and character and (the text omits this) gender. Each child is different from the others, and the same child will be different from year to year, so the ritual presents a series of examples. The first son—the "wise" son—asks, "What are the testimonies, statutes, and judgments which the Lord our God hath commanded you?" The phrasing of the question implies that this son has already studied and understands the richness and complexity of the tradition he is asking about. He asks with considerable maturity, as a child approaching full participation, and the answer he is given, one rather obscure detail of the Passover regulations, evokes by example the whole context of study. The second son is described as wicked or boorish. He is common enough in modern society, rejecting what he sees as meaningless ritual, saying, So what? "What is this to you?" He phrases his question as a nonparticipant, indeed as a nonbeliever, and his father answers in terms that exclude him, "It is because of what the Lord did for *me* when I went forth from Egypt." Here we are, in a text that goes back two thousand years, imagining a son who neither has learned the tradition of his community nor desires to be a participant—a dropout, in fact.

There are two more sons, one who is innocent (often translated "simple") and who receives a simple answer, and one who doesn't even know enough to ask, but is told nonetheless. This fourth son is perhaps an infant, addressed as part of the ritual with information he cannot yet decipher, like little Becky solemnly told again and again that the apple is red. Everyone at the table has been in the position of the fourth son, and everyone is in some sense invited to consider each of the other roles and to strive for the first, the role of the wise and understanding (and studious) son.

At ordinary family meals, in work and play as well as in rituals, the need to sort out who is included and who is excluded is perennial, but there is still ambiguity. In most household activities there is a place for those who are not expert. Even in the most specialized workplaces, in operating rooms and on flight decks, there are differences in kind and degree of expertise. Any human community must find ways to include some who are highly intelligent, some who even as adults barely understand the rules of the culture, and some who rebel or deviate. One of the greatest mistakes made by social scientists is overstating the degree of sameness—of homogeneity—needed for a society to function. Some degree of heterogeneity, if only of age and sex, is necessarily allowed for, and many societies go on to provide niches for visionaries and schizophrenics, the ill and the handicapped, foreigners and visitors.

In societies that do not rely on schooling to educate the young, most learning occurs through participation; it must be true therefore that some participation is possible without knowing the whole complex set of rules. Because we associate learning with schooling, we tend to think of learning as taking place in specialized settings where there is a very clear distinction between a teacher, who knows, and students, who do not, moving through the system in approximately uniform age cohorts and emerging to full participation (commencement) after they have achieved a uniform level of competence. It is on the projected model of this kind of homogeneous adult competence that ethnographers seek out as informants in the field "fully enculturated adults," those paragons who are able to judge whether sentences are grammatical or to tell whether a given marriage is appropriate. The fact that few adults fit this model in our own society has become cause for increasing concern, but the whole notion of homogeneity is questionable. Schools are often modeled on factories, putting children through a production line designed to turn out a uniform product, and we are surprised to find that even so we keep turning out deviants and illiterates—and a few original thinkers. We are increasingly inclined to limit hospitality to newcomers, easily seen as simple or wicked when they fail to conform, or perhaps as undermining the commonality needed for a functioning society.

Does anyone remember Gladly, the cross-eyed bear? She is surely the totem of all children who have sat through adult occasions with only the most partial understanding of the enigmatic behavior of their elders, in this case a Protestant service, the adults singing together, "Gladly the cross I'd bear." We assume that the adults do know what they are saying, more or less, and the children will work it out by the time they are grown up. But grown-ups do not share a complete and unified understanding of the culture of their community. They can participate without doing so, continuing to learn in the process. This is especially true in complex, diverse, and rapidly changing societies, but I believe we have underestimated the extent to which it is true of all human societies. I grew up attending the Episcopal church, sharply aware that under the solemnity of common ceremonial utterances, there was a cacophony—no, a complex polyphony—of diverse belief and understanding. The most beautiful rituals were joint performances sustained by participants with disparate codes. "Jose, can you see" is not a bad anthem for a society of many races and peoples, sung with fervor by thousands who are by no means all Hispanic. Full understanding is not a necessity for participation.

In point of fact, the debate about "multiculturalism" is a good example of an interaction in which the participants do not share the same system of meanings. The term itself is flagged for trouble, like most terms ending in the suffix -ism, immediately identifiable as a matter of contention. No word ending in -ism should be used to refer to the simple fact that the United States is increasingly a society of many races and peoples: motley, dappled, diverse. All that has been said here about communication across difference and the kinds of creativity and insight it can stimulate applies in the United States—as it does in London or Paris or Tel Aviv.

The debate becomes political when questions of policy arise: whether to slow diversification or put a damper on its expression, whether and how to support it. The term *multiculturalism* is used to refer to at least two different but complementary educational strategies: one that supports individuals in their *own* ethnic or racial identities, and one that enhances everyone's capacity to adapt by offering exposure to a variety of other traditions.

I grew up with large doses of Mother Goose, the Five Little Peppers, and Shakespeare, and smaller doses of Balinese music and Russian folktales and the Arabian Nights. By the time I was sixteen I was ready to reverse the proportions, for the mix had somehow made me ready to immerse myself in Hebrew literature and then to move on to Arabic and Persian and Pilipino. When I met my husband I had to look Armenia up in an encyclopedia, but acquiring new skills and information to support the transition from cultural distance to marriage seemed natural. I come from a privileged sector of American society whose identity is affirmed on all sides: at home, in the classroom, and through the media. I was luckier than many WASPs, however, in my early exposures to other traditions, for the privileged are often culture bound, open to only one. Complacency works against curiosity. Minority children, swamped by iterations of Western European traditions they do not identify with, are a lot more canny about the ways of whites than vice versa; street smarts depend on vision at the periphery. Exposure to difference encourages breadth of attention, a way of seeing that underlies ways of continuing to learn as an adult, for every opening to different cultural traditions is a rehearsal for dealing constructively with inner or outer change, adapting to the computer revolution or to one's own aging.

Confidence depends on identity; adaptive insight depends on difference. Efforts to support either through education are referred to as "multiculturalism," which makes the conversation hopelessly confusing. West African legends were for me a part of what might be called "adaptive multiculturalism," but they are a part of "identity multiculturalism" for African American children. Oddly, I probably had better access as a child to these wonderful stories than did black children of my generation, and I was protected from the suggestion that there is only a single way to be human or even to be a good American. I was even luckier in that some of the models I was offered as a child were women of achievement—enough so that there too I could learn there was no single ideal way to be.

Identity multiculturalism is promoted as a way to increase self-esteem, particularly in groups that have been denigrated or discriminated against. The Spanish friars who brought Catholicism to the Philippines offered a place in the vast international community of the church, but it was a second-class place in a community where priests, saints, angels, and even God were clearly European. Until the Filipinos rose against the Spanish, Christianity could only be accepted at a cost to the self, reflected in the mutilated carving on my desk, and modern Filipino nationalism has needed to repair the wounds of the past. Identity is parochial before it is inclusive: for the individual it offers the opportunity to say, This is where I come from, the starting point for knowing who I am, necessarily at times ethnocentric or even separatist. Identity multiculturalism is only *multi-* from the bird's-eye view of planners, who acknowledge the need in many groups. Adaptive multiculturalism, by contrast, is indeed *multi-* for the individual. It is often promoted to increase tolerance and civility, but its greatest importance is in offering multiple ways of looking at the same question.

Between these different kinds of multiculturalism and the many contexts in which they might apply there is plenty of acrimonious and confusing debate about proportions, about how much exposure which child needs to which traditions. Solutions vary from the token inclusion on reading lists of single works by persons of color to the rejection and exclusion from those lists of all works associated with the Western tradition. Most debates fail to recognize that the appropriate proportions vary through the life cycle and that no choice of emphasis is final, for education is never complete.

I like to imagine a campus demonstration in the name of adaptive multiculturalism. A young woman of Chinese ancestry carries a placard demanding tuition in Arabic. A blond young man complains of the absence of courses on Hindu philosophy. An African American protests the lack of a program in Indonesian music. Others want to study Caribbean history, Native American ecology, and gender in Oceania—as well as Emerson and Virginia Woolf and the French Revolution. The students are as diverse as the population, and—amazingly—what they are demanding is not a greater representation of their own traditions in the curriculum but access to the cultures of others.

Disparate as the subjects demanded are, they combine to carry a double message: The first is a moral message of respect that sometimes triggers resistance from students impatient at being told what they ought to feel. But the second message is perhaps the most important and practical message of education today, that because there is no single way to be human, the particular patterns of contemporary American culture should be regarded as mutable. These fantasy demonstrators are demanding that the university broaden their knowledge of the range of human possibility, equipping them to question whatever had been taken as common sense, enhancing the capacity of each to contribute a distinctive point of view. This is essentially a pragmatic demand, not a moral one. Because we cannot tell students how they will have to behave or what they will come to believe in the twenty-first century, we need to give them a variety of examples of what adaptive systems look like, a variety of ways of seeing the world.

This "demonstration" is taking place, quietly, all over the country as students register for courses (and as they tune their radios to "world music"). Its steady influence is masked in the debate by, on the one hand, identity multiculturalism, the understandable demands of many groups that their own traditions be honored and, on the other, the fear that too much diversity may be pathological. This demonstration is not made up of African American students majoring in Black Studies, but of black and white students together taking courses in Chinese, West African, and English cultures. Oddly, the study of history, even that of one's own group, sometimes serves both purposes, since even the near past is exotic for many American youth. I sometimes send my students to interview their own grandparents and find that this can model an encounter with the "other," and African Americans often feel a mixture of pride and alienation when they encounter Africans.

Children develop through concentric worlds, gradually able to move further from home but always seeing each larger sphere through the lens of the previous stage. As a black student of my husband said to him in the sixties, "You have to know where you stand before you can decide who you stand next to." Identity multiculturalism suggests offering multiple tracks at the beginning of education and makes good sense as a place to start, the best place to develop confidence from which to encounter the new or strange, but it should not be exclusive or set up a single axis of contrast, for bicultural contrasts lend themselves to polarization. Black history belongs early in the educational process for African American children, but should not be limited to them, while all children need to be aware of diversity in Africa and in the African diaspora as well as diversity in other shades.

If elementary school were the end of learning, pressure to limit teaching to the "majority" culture would be understandable, but fortunately it is not—not unless the whole process is rejected by children convinced that it offers them nothing. Adaptive multiculturalism has more to say about learning through a lifetime, in a continuing process of encountering difference, steadily going beyond any traditional canon. Traditional liberal education was identity-based for the privileged and identity-threatening for others, yet much of the material covered may be essential for the adaptation of those who have been treated as outsiders.

This is an anxious time in world history, when increases in travel and communication mean that every society must live with difference in a new way. Today, one version of American culture is seen by many as the only basis for national survival, threatened and undermined by multiculturalism. Without a coherent culture, without institutions in their so-called traditional form, we are told, the people will perish. Young people exposed to variety may risk becoming confused or alienated, and fundamentalists object even to the teaching of classical mythology in the schools. Enhancing identity, increasing awareness of difference, we are genuinely playing with fire, for the resurgence of repressed nationalisms can destroy societies, and almost any kind of difference can provide a basis for discrimination or exploitation. But although no one knows the minimum level of commonality for cultural viability or how to use the models offered by the many multiculturalisms of human history, we do know that difference stimulates creativity. When we talk about going beyond the traditional canon, we are talking about opening up a library not of great books but of versions of humanness—some of them never written down in any form at all, but many of them in written form, often in rather surprising places.

Even the texts of orthodoxy can be taught so as to open windows to diversity, revealing the different, even the exotic. In the first century of the common era, the rabbis were sorting out which texts should be included in the anthology that is today the Jewish Bible and the Old Testament for Christians. Part of the fascination of the Talmud is that it is a record of constant argumentation, of a style of reasoning that mixes logic with metaphor and legend, suggesting in its own way the value of dissent. Debate focused on two books that seemed to some downright ungodly, the Song of Songs and Ecclesiastes. These two books are attributed to King

Solomon but associated with different stages of his life, for the Song of Songs expresses the erotic love of youth and Ecclesiastes the decrepitude and despair of old age. In making a case for including both, Rabbi Akiva argued that the Song of Songs is not about physical love but is an allegory of God's love for his people. This made it respectable—and ethnocentric, though the metaphor has been easily carried over into Christian ideas of the church as the bride of Christ. Whatever the arguments, they had the effect of preserving two extraordinary bodies of poetry and making sure that the Biblical canon included a record of sexual passion and of the alienation of old age, experiences that bridge many cultural differences. Their inclusion in the Bible, along with many less than edifying historical narratives, has meant that in tens of thousands of households children growing up with strictly censored reading could pass from one to another the guidelines to a somewhat broader education. The great poem in the third chapter of Ecclesiastes, "To everything there is a season, and a time to every purpose under the heaven," might even be taken as a sort of rough index to a canon of human experience, every chapter of which is interpreted differently in different times and places, and some of which are excluded or denied. Sex. Birth. Dying.

We possess after all the essential basis of commonality. Of all the texts that must be read to understand the human condition, the body is the most eloquent, for we read in all its stages and transitions a pattern that connects all human communities as well as differences that divide. People in different eras and places have read it differently, or made every effort to deny access to parts of the story, to its alternate readings, or to the wider learning that flows from it, so it becomes the justification for mutual suspicion and for alienation from the natural world.

The Bible can serve another purpose today that is equally likely to be obscured, which is to provide contact with a different and alien culture. Damavand College, where I taught for a time in Iran, was derived from Presbyterian mission schools, and although it was formally nonsectarian its requirements might have been seen as preparations for evangelization. The basic concept of the core curriculum came from the Great Books curriculum of Robert Maynard Hutchins, with certain key changes. On the one hand, a set of Iranian and Islamic Great Books had been added. To balance this, the list had been trimmed down to reflect only one side of the Western tradition, the side that leads from Plato through Augustine and up to Pascal. The other side, which winds its way from Aristotle through Aquinas, was scanted, and the Great Books of natural science were removed. Science was represented by a single required course in biology; social science was represented by one course in anthropology and a variety of electives. The planners had removed many of the great debates of Western civilization that keep us thinking, but they had replaced them with the encounter of two cultures. The total had to be modest, for except in Islamic and Iranian subjects the students were reading in English, a second language for them.

This curriculum gave me a rich reserve of case material for anthropology, for instead of presenting the classics as evidence of continuity in Western civilization, I could present them as evidence of the human range of variation. The Book of Ruth, so often read as a testimony of love and fidelity, proposes the idea of marriage not between a man and a woman but between a family and a woman, including the custom of the levirate, giving a widow to some other male in the husband's clan to make a child in his name. In my classroom, the Greek polis illustrated not the various forms of government we know today but forms of government in face-to-face communities, tribal warfare, the overlay of patrilineal patterns upon matriliny. Text after text provided a context for obliquely discussing the institution of monarchy in a setting where explicit discussion of the Pahlavi dynasty and its doings was closely monitored. When students read the Gospels as texts describing Middle Eastern culture patterns, they could see them not as the foundations of foreign—Western and Christian—traditions but as local, acted out in contemporary behavior. At an Iranian dinner party, the room always has a high and a low end, and the appropriately modest guest will seat himself at the bottom of the room until the host calls him, Come up beside me . . . just as Jesus advised.

Setting material about other cultures in the context of

the life cycle also makes empathy easier, yet we do not read the texts of the human life cycle directly, just as we cannot read written texts from the past as their writers intended, for experience is always filtered through previous experience, and always interpreted in cultural terms. It is only occasionally that we are reminded of how indirect most experience is. And it is even more rarely that we are reminded to accept responsibility for the way experience is constructed.

Participation in interactions in which codes are not fully shared is not the exception but the rule, but life goes on. Theories of human social behavior and adaptation that obscure how often communication is communication across difference are like descriptions of the bumblebee, declared by the aeronautical engineer to be unable to fly. Everybody speaks a slightly different version of any "common" language (what linguists call an idiolect). In fact, one reason that it has become fashionable to think of language as innate is the observed diversity and irregularity of the speech environment children learn in—yet learn they do, less distressed by a variable reality than the linguist is by his theory. The visiting ethnographer, who feels like odd person out, projects a uniformity on the "natives" that complements her or his own sense of isolation. This is amplified by generalizations used in description, and further nuances are lost in popularization.

When we think of communication and failures of communication across difference in this society, we tend to think in terms of social problems surrounded by rhetoric, particularly of racial or interethnic conflict. Gender, as described by Deborah Tannen and others, offers a slightly better model of communication across difference, contested but ongoing, since societies can split on ethnic lines but never entirely split on gender, in spite of misunderstandings. No solution to these issues is going to be perfect: there is too much accumulated discomfort, too many bruises, so progress alternates with phases of exaggeration and backlash.

Much of the difference within any society has to do with where individuals are in the life cycle, which everywhere involves ongoing learning. Thus, an even better paradigm than gender might be those interactions between parent and child starting within weeks of birth, in which, although it is obvious that the child has not yet mastered the codes and patterns of the culture, the two learn how to sustain the joint performances that apparently provide contexts for learning. In the United States, different generations have matured under such different conditions that cross-generational communication is almost as problematic as interethnic communication. Notoriously, the grandchildren of immigrants are the ones who pull back from parental assimilation and seek their roots.

The distinction between the process whereby children learn their "own" culture and the process of exposure to and influence by a foreign culture fades in a diverse society that depends on lifelong learning, in which every group is necessarily diverse. Thus, learning through the life cycle offers a model for studying internal diversity even in relatively stable and isolated communities. Participation often involves skills in coping with ambiguity. "Common sense" is often assumed to be truly common and to precede specialization, but it may be that "common sense" is mastered only in old age, when it is revered as "wisdom."

One conspicuous strand of contemporary debate attempts to inventory what every member of society needs to know, whether in curricula and proposed standard examinations, in such more fanciful forms as E. D. Hirsch, Jr.'s *Cultural Literacy*, or in the so-called canon. No one, it might be argued, is a full participant in American society who is not numerate and literate in English, does not know enough of the rules of baseball and civics to take sides, and so on and so forth, perhaps at very great length. Depending on how we define *full participant*, it may be essential to have read Melville or to understand the theory of relativity. It may also be necessary to know how to program a VCR or how to fill out an application for food stamps. No one, it might be argued, is a full participant in American society who does not have some basic knowledge of the histories and folkways of the diverse groups that compose that society. Some knowledge of Buddhism and some of Vodun. But are there any competent participants in American society? Young people must be prepared to feel like newly arrived immigrants through much of their lives. They need to know how to observe, how to learn, how to adapt, how to draw on other people's expertise. How to improvise and cope with only partial knowledge and how to imagine alternatives.

On the one hand, we overestimate the extent to which cultural patterns are shared and regular. But, on the other hand, we underestimate the possibility of getting along without complete sharing, and the self-organizing properties of human communication. We assume that patterning must come from the past, underestimating the way in which a group of strangers can learn not only to interact in harmony but also to generate new regularities, which come to be treasured. The current debate about how much cultural pluralism any society can tolerate is partly a matter of growing pains but also a symptom of the need to rethink the relationship between identity and adaptation. On the one hand, the demand for multiculturalism is a demand for acknowledgment, a recognition owed to every member of society as a foundation for the sense of self that will allow ongoing learning. But, on the other, multiculturalism fits with the promise of lifelong learning, the promise that individuals will increasingly be able to look at one another with recognition as they continue to grow out of fixed roles.

Educators, even in universities, are still strangely in thrall to the idea that education *precedes* participation, even though more and more adults are returning to study or working and studying simultaneously. Education is less and less a preparation for life and more and more a part of it. Except in a few privileged colleges, diversity of age and experience has become the most important resource of the teacher, balanced and supplemented by diversity of ethnic background, so that classrooms can increasingly be orchestrated in ways that allow students to benefit from collaboration and teach one another, mining difference for insight.

In ecology, diversity may offer resilience to systems, whereas large plantations of single crops are highly vulnerable to pests and diseases. Because of the increasing concern that cultural diversity might prove toxic to society, I was intrigued to learn that diversity could be used to reduce toxicity. Areas of extreme pollution are sometimes seen as emblematic of wider problems in society and in the biosphere, like the famous Love Canal. In the south, there is a five-and-a-half-mile stretch of Chattanooga Creek polluted with a wide range of toxic substances from forty-two hazardous waste dumps. Those who live nearby, who are mainly poor and black, complain of high rates of leukemia and other cancers. Running a mere hundred feet from a school yard, this creek has sediments eight feet thick, containing the insoluble large molecules of polyaromatic hydrocarbons, coal tar derivatives, and pesticides. The usual Superfund tactics for solving such problems are dredging and incineration followed by containment, for although progress has been made in using biological systems, such as pools filled with water hyacinths, to process ordinary sewage, these systems have on the whole been kept deliberately simple so that they would be predictable. It makes engineers nervous to think about more than one organism at a time, for who knows what unexpected patterns might emerge from interactions?

John Todd of the Center for the Restoration of Waters described to me his experiments with techniques of water purification using biological diversity. He has been interested over the years in human habitations that are ecologically benign and self-maintaining, and subsequently in using biological processes for repair, calling his constructions living machines. What John does is maximize diversity, particularly at the microbial level, where sunlight and photosynthesis can be brought to bear by busy communities of bacteria, but also including the full range of organisms from bacteria to vertebrates. He seeks out organisms that have survived in highly stressed environments and makes sure that his living machines involve recurrent sharp contrasts. Polluted waters are passed through enclosed tanks, then through tanks in greenhouses, and finally through outdoor marshes. The opposite of stagnant ponds, where heavy metals sink and pollutants separate in layers, his processing tanks constantly bring different substances and organisms into contact with one another, a kind of conversation.

This work is still highly experimental and hard to fund. John claims to have removed more than two-thirds of the different toxins almost entirely and others partially, much more cheaply than by conventional methods, but this may not be enough. What is interesting (and makes engineers nervous) is that no one knows how to choose all the organisms to participate in such systems, so John gathers samples from swamps and rivers and puts them together, allowing them to organize their own communities. They are too complex to plan, just as no available process of deliberate planning could generate the cultural diversity of a great city.

The preference for uniformity has deep roots. Agribusiness still prefers to work with monocultures, thousands of acres of uniform planting requiring heavy use of pesticides and fertilizers. Contemporary anxiety about multiculturalism may echo recurrent attempts to root out heresy and maintain spiritual monocultures. Yet oddly, once we accept the presence of diversity, it becomes possible to find it in new forms, while at the same time subtle echoes become discernible in what once seemed entirely disparate. Internally as well, diversity and congruity can combine in the liberation of unimagined potentials.

12

Limited Good

WHEN STRANGERS MEET AND TRY TO FIND WAYS of fitting together, the outcome may depend on the assumptions they bring to all interactions, even before the first new cue is available. In the societies I have studied, some of the ways of seeing the world go back to common sources thousands of years old, for in each of these societies related religions provide recurrent metaphors. Fathers and the authority of fathers. Sacrifice and law. The divine right to dominion over children, women, other species, other nations, the planet itself.

Judaism, Christianity, and Islam have been shaped by a recurrent solution to encounters with other ways of seeing the world: whether in the family or in the universe, authority has a single source and so does truth. Abraham's God is a jealous god, and truth is exclusive. Each community believes that its understanding is not simply good, not even better, but best. We cannot all be right, so whatever is different is wrong. Here, identity precludes adaptation. Heresy is another of the themes of this tradition, carried over in some of its nontheist offshoots, like psychoanalysis and Marxism.

These ideas are still being acted out in the Middle East, reemerging in the former Soviet Union and Yugoslavia, inspiring civil wars in Africa. They were expressed in the human constellation joined to sacrifice a sheep in that Persian garden and in the Islamic revolution less than a decade later. In the Philippines they are filtered through generations of missionizing and syncretism, but they still cause friction between different kinds of Christians and between Christians and Muslims, and scorn for the pre-Christian beliefs that survive in the mountains.

The same symbols turn up in different traditions. There is an egg on the Seder table, and eggs are part of Easter celebrations from Washington, DC, to Manila to Madrid. The eggs on the festive tables celebrating the vernal equinox in Iran are pre-Islamic, going back before Zoroaster, suggesting that the common symbol is even more ancient. An egg appears to be dead, like a stone, but brings forth life.

Many of the same basic ideas about human use of the natural world also turn up in Judaism, Christianity, and Islam. In all three, men (in the old ambiguous sense with its hidden preference) stand in a special relationship to God that separates them from the rest of creation and makes them lords over it. Ideas of what that creation is like—knowable, divisible, subject to use and domination—also turn up again and again. The wooden Christ on my desk with its nose sliced off reflects the perception by Filipinos, a Malayo-Polynesian people, that there was a connection between Spanish domination and the imagery of Christianity.

The Book of Genesis is a compendium of origin myths including the first incest and the first refugee as well as the creation and the fall and the first murder; running through it there are echoes of other ways of seeing the world and the gradual assertion of a new point of view about both ideas and real estate. Abraham's grandson Jacob appears as one of those culture heroes, like Raven or Coyote or Odysseus, celebrated for subtlety and guile. I like to think of him, when he cheats his brother, as the inventor of the world of zero-sum economics: the world where every success is at the expense of someone else.

Old as the story is, it is still with us, and so is the seemingly self-evident truth it embodies. Every classroom in which grading is done on a curve (as it typically is in the large low-level economics classes that students flock to) reinforces the message. There are only so many good grades, and every gain for someone else is a loss for you. The tests and exams enforce a view of the world more profoundly than the lectures do. The question of when and whether people are able to learn new views of the world should not be graded on a curve, however, for knowledge is like the Biblical loaves and fishes, increasing when it is shared. The principles that turn up in grading and electoral politics, above all in our use of that fluid resource called money, are based on the metaphor of arable land and are still fought out over land in the Middle East.

Abraham had been a migrant herder, seeking land and water. The full transition to farming was made by his son Isaac. When Isaac found a place to dig a well that was not disputed by earlier inhabitants, he celebrated its amplitude, saying, "Now the Lord hath made room for us and we shall be fruitful in the land" (Genesis 26:22). But settled agriculture is a mixed blessing, bringing a change in the nature of property and thus a change in relationships. Isaac's son Jacob, who was a farmer, set out to defraud his brother, Esau, who was a hunter, recognizing that with finite resources of arable land, competition would replace sharing. First he persuaded Esau to give up his birthright for bread and a porridge of red lentils, cultivated food. This famous meal is one of those staples of farming societies, key to the Neolithic revolution, that provide complete proteins by combining grains and legumes. Later Jacob deceived his father into giving him the firstborn's blessing as well as the inheritance, by offering his father the meat of a domesticated goat while Esau was out hunting.

When Isaac blessed Jacob, he blessed him with the benefits of agriculture—"the dew of heaven and the fatness of the earth and plenty of corn and wine." He also blessed him with dominance, saying, "Let people serve thee, and nations bow down to thee: be lord over thy brethren, and let thy mother's sons bow down to thee" (Genesis 27:28–29). Almost unnoticed here, the struggle for ascendancy replaces and subsumes the struggle for limited resources. The world of zero-sum economics tends to become a world of some people ruling others. Hunter-gatherers share food with kin, and herders like Abraham depend on a form of wealth that is easily subdivided and moved, flocks that will increase given care and good fortune. But settled farmers depend on the ownership of limited areas of arable land that cannot be efficiently farmed if they are divided up too often.

Origin myths do not of course explain how ideas and institutions come into being, but they reflect the presence of those ideas and reinforce them generation after generation. The ideas conveyed in these Biblical tales pervade the cultures derived from Abraham's tradition, but they have also developed in other cultures committed to the combination of long-term settled agriculture, patrilineal inheritance, and increasing numbers of progeny. Assumptions run deep, hard to recognize and hard to change, setting the terms of many interactions by posing the question Who in this encounter will win? The calculus of game theory, as described by John von Neumann and Oskar Morgenstern, has a certain elegant simplicity that avoids ambiguity.

Before learning to grow their own food, human beings were far and few between, dying early, following the game, and moving in response to changes in climate. Then, some twelve thousand years ago, with the invention of agriculture, populations began a slow and steady increase, setting the stage for new conquests and migrations as communities grew toward and passed carrying capacity. New technologies were developed, some that made warfare more efficient and others that made it possible for more human beings to survive on the same territory.

Many peoples see the world in terms of what the anthropologist George Foster called the Image of Limited Good, which is the lesson often drawn from limited and repeatedly subdivided arable land: the pie is only so large, and an increase in my neighbor's share means a decrease in mine. The Fertile Crescent was one of the early centers of agriculture and population growth, accelerated by the invention of irrigation and the rise of empires and great cities. One man's loss came to be another man's gain. The myth of Jacob suggests that he understood this and used it to his profit.

The life of hunter-gatherers like the San does not generally fit this pattern. These are societies in which property is scant and easily portable, the sharing of food is obligatory and enforced by social pressure, and competition is not emphasized. No group is without conflict, but a number of hunting-gathering peoples lack the institution of warfare. Even where warfare is present, it is striking how often warfare among preliterate peoples is not organized around territory or sustained dominance but around ideas like honor. Modern warfare, however, behind its complex ideological overlays, has been very greatly concerned with territory and control of resources, oil reserves or deepwater harbors. The end of the cold war removed an ideological polarization but unleashed the rivalries within the Soviet empire, the enmities between brothers when the father's control is removed.

When Europeans spilled out of their intensively farmed enclave in Europe, they found their way to lands where subsistence patterns spread people farther and densities were lower. Most of Australia and Oceania, parts of the rain forest areas of Southeast Asia, much of sub-Saharan Africa and the new world depended on hunting and gathering or on shifting horticulture, in which a household could increase its food supply by opening up new fields. In all these places, zero-sum economics was introduced as a benefit of civilization, for often new ideas of salvation and new attitudes toward land have come in the same package.

A change in the distribution of humans on the land was well under way when Europeans got their first toehold at the southern tip of Africa. Farther north, a number of peoples speaking languages of the Bantu family had been gradually shifting their herds and farms southward, bringing new areas under cultivation and displacing hunter-gatherers. Both the Boers and the English came to Africa bringing European ideas of ownership and competition, and the Boers especially brought imaginations dominated by Old Testament imagery of a chosen people and a promised land. In Africa as in North America, resources that once seemed limitless were made finite, boundaries were drawn and fences built. Eventually apartheid laws were passed, limiting the access of black Africans to land and later segregating them in pseudostates, the national homelands, with desperately inadequate resources for growing populations. A few generations of European colonization ground in the lesson of limited good and ground down the ethos of sharing, so that the abolition of apartheid immediately posed the question of succession, of larger slices of the newly available pie. Today, the fault lines in South Africa recapitulate global problems, the interface within a single country of the developing and the industrialized world.

Land and water are not the only kinds of benefits that people came to imagine as divided in fixed shares so that more for one meant less for the other. Isaac's paternal blessing is as much a matter of envy as his wealth. The rivalry of two brothers vying for a father's love is prefigured in the rivalry of Cain and Abel for God's favor, which seems to echo a time very early in the history of the Jewish people when the life of nomadic herdsmen still seemed morally superior to that of farmers. One brother offered God meat while the other offered fruit and vegetables and grain. God favored the offering of Abel over that of Cain, who in his jealousy committed the first murder, killing Abel. Modern readers tend to be disconcerted by the fact that Cain, the murderer, was the farmer, the one who offered veggies to a God who preferred meat. The hostility between Jews and Arabs is traced to another pair of brothers, Ishmael and Isaac, when Sarah decided that marriage was a zero-sum game and Ishmael, with his mother, Abraham's concubine Hagar, was abandoned to God's care.

In such a worldview, blessings are limited. Love is limited. Not only are all goods limited but they are all fungible and all transitive, so that it is easy to pit different values against each other, as if they belonged to the same kind of arithmetic, spotted owls versus full employment. For many doctrines, salvation is limited and there will be wailing and gnashing of teeth. Listening to a Native American prophet speak of the doom to be visited on Euro-Americans, my father leaned over and sang to me, *sub voce*, "The bells of Hell go tingalingaling, for you and not for me . . . " A very popular idea.

There are other visions, however, which prosper in times of expansion and new technology, and which suggest that prosperity is not limited and neither is love or blessing. In the contemporary world, we restate these conflicting views in the ongoing debate about whether technology will always be able to refill the cornucopia emptied by human prodigality, or whether we must indeed accept "limits to growth." Yet economic and population growth are not the only kinds of growth, for lives can be enriched and deepened in ways that do not exclude or strain limited resources.

The stands that people take on these issues are not based on rational analysis, they are based on very early learning of how the world really is, reinforced by teaching of other kinds. It is impossible to avoid some rivalry between siblings, just as it is impossible to eliminate completely the themes of dominance and dependence that all human beings must deal with because we are born so small and weak, so dependent on larger and stronger beings. Today American thinking on child rearing urges parents to avoid invidious comparison between siblings and suggests patterns that promise enough for all and a fair division, but competition is affirmed at school. Because this kind of calculus is so simple, it tends to force out other ways of thinking, particularly in the ambiguous encounters of strangers. It provides a map, ultimately a destructive one, for finding the way through interactions in which no code is shared.

Middle Eastern cultures reek of envy. Everywhere that you see charms against the evil eye—bright blue beads, the hand of Fatima, an eye painted on a house—you know that you are in a society where envy is a major source of anxiety. The evil eye is caused by envy, often involuntarily. If you admire the fat and healthy infant of your neighbor, that baby may sicken and die, so in voicing admiration you should always invoke a protective charm: heaven forfend that he should come to any harm (the admired baby is almost certainly male—female infants are not enviable). A baby is protected from birth by charms attached to clothing or body, kohl on the eyelids. Concealment is another sign of an envy culture, as in Iran, where women are veiled and gardens surrounded by high walls to protect them from covetous eyes. The flashy houses built by the nouveaux riches under the shah were an innovation, for the wise man does not flaunt his wealth. The traditional merchants of the bazaar hid their money away and kept their luxuries invisible.

A few coins to a beggar are a protection against the evil eye, but serious efforts at social programs are often shipwrecked on disagreements between those who feel that everyone will be better off when no one is starving and those who feel that every penny going to the poor will come from their own pockets. When the economic shoe pinches, increasing numbers begin to grudge any advancement to their neighbors, even welfare payments. In many countries today there is a curious tension between the effort to keep competition for wealth alive, as a carrot to motivate economic enterprise, and the effort to establish a basic standard of welfare, eliminating the stick.

The Ottoman Empire represented long-term neglect and deterioration, with the result that populations in many areas, unprotected from nomadic raiders, declined. Systems of irrigation and drainage were allowed to run down, so land that was once cultivated was given over to wilderness. When Jews first began to immigrate into Palestine, some set out to buy land, sometimes from owners living on it but often from absentee landlords who did not care about the displacement of their tenants. During the British mandate, legislation was passed to protect tenants, so that often the only land settlers were able to buy was uncultivated (although generally it was used for some other purpose, like grazing or gathering firewood). The Jews were legalistic about ownership: what they had paid for was theirs. No doubt the purchasers of Manhattan felt the same way. Although the new settlers were initially inept farmers, they soon began to bring in know-how and farm more productively. There are many stories in Israeli lore of the kindness of Arab neighbors to the early arrivals, of hospitality and skills shared, but this kindness evaporated when Jewish farms began to be more productive than Arab farms.

Since this period, land has been taken in battle and been expropriated for settlement and for defense, but it is important to understand that the hostility that developed in the years leading up to Israel's independence is explainable not in terms of details of how each particular piece of land changed hands but in terms of differing understandings. According to the theory of limited good, when the settlers prospered this could only be at the expense of their neighbors. For all the years of Arab-Israeli conflict, observers

have been pointing out that the highly educated and entre-
preneurial Jewish state could become a source of prosperity
for the whole region, but this idea is implausible to those
who are convinced that all prosperity is at someone's
expense. Good is limited. Agriculture was after all invented
in the Middle East. Premises so deeply learned are confirmed
by experience, for contrary evidence is simply invisible.

Some people, on the other hand, genuinely believe that
peaceful solutions to conflict will increase the total of available benefits, for peace and freedom from fear are among
those goods that need not be limited. In South Africa progressive whites eventually realized that apartheid was hin-
dering development even before the international boycott,
and saw that the nation's prosperity was an illusion based
not only on an unequal division but on an extractive economy: wealth mined from the ground rather than produced by
human effort and ingenuity. The positive vision that combined with fear in persuading the government to start the
long, slow transition to equality was the vision of an educated workforce becoming increasingly productive and inno-
vative in a world of high technology: more for everyone.
Such a workforce can only be created with a degree of social
equality. Education is one of those goods that is most clearly
not limited, so white South Africans are more ready to commit themselves to raising educational levels among black
South Africans than to commit themselves to land reform,
looking ahead to an industrialized economy in which only a
few live on the land and productivity depends on technological development. But assumptions once learned are not so
easily left behind, and generations of deprivation have meant
that the promise of increased prosperity—a new pie to be
divided—triggered rivalries within the black community.

In the Philippines, people often choose to mute the
appearance of competition. If you comment on the success
of one member of a team or one child in a classroom, the
immediate explanation is that she or he succeeded by chance
rather than superior skill or intelligence. There are traditional contexts of common effort, like preparations for community celebrations, that thrive on this approach, which is
undermined in foreign models.

In Iran, there is no reluctance to profit at the expense of others, but it is necessary to be circumspect, for conspicuous success immediately attracts suspicion. A friend once remarked, in a common Iranian slur, that the prime minister was undoubtedly a passive homosexual, probably because of being sodomized as a child. "How on earth could you know that?" I asked. "Look," he explained, "the man could not possibly have gotten where he has unless he were without honor, ready to engage in any possible corruption. He must have lost his honor early on so that now he can be without shame and do anything he needs to do to get ahead." QED. My friend, a highly educated man who had spent many years abroad, felt disabled by his own principles and was convinced, as he looked around him, that the success of others must be dishonorable. This line of thought fuels a desire to discredit and topple those in power, followed by a recurrence of suspicion of their replacements. Not only is good limited but those who succeed must have done so dishonestly, and all victors are suspected of having won by cheating. Both views conflict with the possibility that everyone benefits, basking in reflected light, when "one of our boys" makes it.

It is hard to sort out the rights and wrongs of attitudes that are so fundamental. Landownership has often been a genuinely zero-sum game, and often the only game in town. There are some goods that are very clearly finite and others that are flexible, and we differ in the goods we choose as models. For some, wealth shared with others loses its savor, and want becomes doubly bitter. They act, even in the classroom, as if every ounce of understanding could become someone's property. Instructors reinforce this approach to learning by their methods of grading, and medical schools virtually require it of their applicants.

The opposite of a zero-sum game is one in which the prize is not fixed. Instead, it grows if the players agree to share it, and may even disappear if they do not. Those who are preoccupied with competition, with doing better than others, are the least likely to thrive in such a climate.

Cultures often seem to allow for opposites in one form or another, however, especially in complex societies, and themes of cooperation coexist with themes of competition. Side by side with the most competitive of premeds, there are students who recognize the world of knowledge as a place with room for all. The most fruitful innovation in education may prove to be a new emphasis on collaborative learning at every level. Knowledge, however much we try to assess and protect it, is potentially at the opposite extreme from land: too narrowly preserved, it may be lost, like many secrets of medieval craftsmen; widely shared and debated, it is likely to grow. Science is not only a method of discovery and verification, it is a pattern of sharing knowledge.

Modern thinking about evolution has made it clear that survival is not a simple zero-sum game; the new inference is that unless we share the planet with other species we imperil our own survival. Competition has its uses but often failures of cooperation lead to mutual loss: a factory that goes out of business because no compromise is found to end a strike, putting all the workers on the street, or nations that are incapacitated by mutually destructive warfare. The belief in limited good may lead a man to sabotage a neighbor out of the intuition that the neighbor's success will somehow be his loss, and simplistic interpretations of the survival of the fittest will lead him to watch his neighbor's failure without concern and without empathy. But the recognition of a common good may lead to preventing a neighbor's failure, in the belief that the entire community will benefit. In international politics, if warfare is the cost of error, then diplomacy is not finally a zero-sum game. Even in free-enterprise systems, government must occasionally rescue failing industries, as in the Chrysler bailout.

Traditionally in American society, men have been trained for both competition and teamwork through sports, while women have been reared to merge their welfare with that of the family, with fewer opportunities for either independence or other team identifications, and fewer challenges to direct competition. In effect, women have been circumscribed within that unit where the benefit of one is most easily believed to be the benefit of all.

Noncompetitive strategies are more likely to be adopted when people know one another. But "knowing one another" can take many forms, and alliances come into being for many reasons: alums of the same college, even if they have never met. Members of the same regiment. Homeboys. Trust and cooperation are created symbolically and affirmed by such mechanisms as Israel's Law of Return, which asserts that every Jew has a right to come home to Israel. In many settings where people do not really know one another, they act as if they did. The enmities or affinities that become the basis for cooperation or competition are socially constructed. The famous distinction between a house and a home acknowledges that home is an idea, a belief, a context for sharing that may be as large as a nation or a planet, yet in some traditions cooperation thrives only against a background of enmity, as in the often quoted Arabic proverb "My brother and I will fight my cousin; my cousin and I will fight the stranger."

The logic of twelve thousand years of agriculture is now obsolete as a model for social life in developed countries, where fewer and fewer work or live on the land and food production is no longer labor-intensive. The problem we face is not the limitation on the supply of land but the lack of jobs; yet jobs can be created, as they have been over the past century, absorbing a vast outflow of labor from agriculture. New jobs depend on the imagination and prosperity of the community. Money and credit are both artifacts of fragile belief and capable of stretching or contracting.

In effect, the Image of Limited Good, which was at least partly supported by experience, is no longer true except to the extent we insist on making it so, but it continues to be central to the experience of third world nations. It can only be left behind if we succeed both in leveling off population growth worldwide and in building institutions to reflect nonzero-sum patterns of thought. The discovery on that hike in the Sinai, when I virtually had to be carried by my companions, that what I would have resented at home could be treasured and built upon there gave me a way of understanding social life in which individuals were not defined as losers or winners at one another's expense. There are many contexts in which it is important to interpret even failure and defeat as part of a valued effort that is beneficial to the whole.

American society loses immensely by its increasing interpretation of electoral politics in simple win-lose terms. Some
campaigns develop the understanding of issues and agenda
for the future, providing a broad social benefit, while others
degrade the process so that the winner, like so many victors
in warfare, conquers a bitter and burnt-over community. The
electoral process was designed to create a forum for discussion in which honorable men argued different views and
made their choices with mutual respect, not to turn half of
all participants into losers, shamed and burdened by debt,
yet the process today favors the entry into politics of people
who think in zero-sum terms. Only when potential candi-
dates can see their participation as valuable, even when they
are defeated, will we have men and women in office who can
be concerned for the common good rather than for victory in
the next election. My suspicion is that it is exactly this way of
formatting the political process that makes it so uninviting to
many men and perhaps most women, for neither victory nor
defeat seems to offer them what they most value. We live
with a tension created by building pluralism on an ancient
tradition of exclusive truth.

The most important arena in which all of us have to
operate by a nonzero-sum logic is the environment, sometimes seen as a player and sometimes as a prize. Measures
taken to protect the natural world often pose dilemmas like
those elaborated in game theory with the classic story called
prisoner's dilemma, in which two suspects for armed robbery
are imprisoned separately. If neither confesses, they will both
have minimal sentences, simply for carrying weapons. If one
turns state's evidence, he will go free and the other will be in
jail for many years. If both confess, both will have jail terms,
although slightly shorter ones. Self-interest and suspicion are
pitted against trust and the common good. Cooperative
strategies often involve smaller gains for each player than the
gains of the winner who plays only to win, for cooperation
only looks appealing if one adds up the benefit to all participants. Similarly, there are costs in reducing pollution, but if
every company in an industry shares those costs, everyone
will share the benefit. When some break the rules, however,
they may get a jackpot while those who accept them suffer.

Every time I accept some small cost or inconvenience to protect the natural world, I am aware that my action not only is
ineffectual but works against me—unless everyone else joins
in. We need to develop structures that favor cooperation
without losing the benefits of competition. It is a matter of
institutional design whether to favor those willing to over-
power the weak or those who advocate ethical behavior and
then cheat.

Experiments with games modeled on prisoner's dilemma
suggest that repetition and face-to-face familiarity are helpful if players are to learn cooperative strategies. This is very
bad news from the game theorists, because as societies grow
larger and increasingly complex, cooperation becomes
harder. Furthermore, there is a substantial philosophical tradition suggesting that cooperation is somehow unnatural,
that nature is red in tooth and claw and the natural relationship of human beings is uncontrolled conflict. With all those
stories about enmity between brothers, it is hard to hope for
cooperation and mutual help between strangers, yet it does
occur, and kinship is an idea as well as a fact of genetics.

All too often all the players share a value system that
works against cooperation—they cooperate in noncoopera-
tion. The compromise that saves a factory employing many
hundreds of workers, the conscientious compliance with regulations to avoid pollution that passes up opportunities for
quick profit—these are not regarded as achievements, yet
they are examples of an essential kind of heroism. All too
many of the achievements we celebrate are won by convert-
ing an opportunity for cooperation into an opportunity for
competition, won by forcing an increasingly interdependent
world into the restrictive mold of conflict.

We know now that the natural world is not only a world
of competition, for the creature that destroys its environment destroys itself, yet we continue to celebrate the
autonomous hero who battles with his fellows and with the
natural world at everyone else's expense. He is visible, easy to
focus on at the expense of family, community, and natural
world; but the welfare of an individual is not bounded by his
skin. It is possible, on the one hand, to go up the scale and
identify one's own welfare with that of a community or even

with the planet, or, on the other, to go down the scale to study zero-sum gaming at the level of molecules rather than individuals, as sociobiologists do.

It would be interesting to review a range of celebrated achievements and valued activities, asking which ones consist in competition for a share in a finite pie and which involve enlarging that pie—enriching the earth. The true achievers could be recognized as those who do not primarily compete or deplete but instead add to the store of understanding or beauty, increase environmental diversity or the human capacity for cooperation. Only a handful of statesmen would fit that definition of achievement: the peacemakers and builders of new institutions, like the social security system. Many artists and scientists. Many entrepreneurs who discover new niches of possibility. And all those who as parents or lovers, gardeners or healers, make it possible for others to flourish. Action taken to permit diversity enriches the earth.

Even truth is not always a zero-sum game, for although some kinds of propositions are mutually exclusive, the many truths of faith and imagination could flourish side by side. "There are no whole truths," Alfred North Whitehead said. "All truths are half truths. It is trying to treat them as whole truths that plays the devil."

The planet may be the final test of whether we prefer competition or cooperation, for the earth is a home we share with many species, not an asset to be divided up among the human players alone. From one point of view, the ecosystem is a player to which, if we try to defeat it, we will lose, but with which we can cooperate in sustainable systems. We treat the planet as a rival when we speak of struggling against natural forces, or dominating nature, but we could learn to treat it as a lover or a child. Imagery of cooperation with nature is scarce in the arid regions that gave us our religions and our history of warfare and exclusive truths, but wonder is also a resource. All three traditions value charity and hospitality, all three work with caretaking and stewardship.

Centuries in which land has been treasured sometimes create landscapes of extraordinary beauty and human skill in caring for them. In mountainous areas in northern Luzon, whole mountain slopes have been terraced for rice paddies, to bring as much land under cultivation as possible; when the paddies are flooded the mountains look like faceted jewels. Parts of this land are so fertile that fence posts sprout into trees, yet the modern history of the Philippines has been a history of the concentration of landownership in the hands of the friars or of an emerging aristocracy, so that most of the population was landless and poor. In Israel, when you walk across the arid landscapes, every hillside is experienced as a text of common history. Hunters in Africa or the Australian outback have studied the land they hunt across and peopled it with legends. But in times of migration or colonization, land conquered first by invasion and then by the plow may finally become an impersonal commodity.

It is curious that the sins of disobedience against God have been emphasized so much more than the sins of hatred between brothers. The story of Jacob's household goes on at some length. He eventually has twelve sons, which might be taken to symbolize the first population explosion, accounting for the origins of twelve different tribes. More sibling rivalry leads to Joseph being sold into slavery in Egypt, where he becomes a visionary and gains great influence. Famine follows, and with Joseph's help all twelve households settle in Egypt, continuing to be so prolific that over time they are reduced to slavery, and then Pharaoh goes on to command the murder of all male infants. A story of conflict between brothers becomes a story of conflict between nations, but both conflicts grow from the conviction that behavior is driven by scarcity.

13

Learning as Coming Home

I HAVE BEEN INVOLVED in one way or another in the educational system of each of the countries where I have lived. In Israel, at sixteen, I learned Hebrew in order to join a high school class and take the national matriculation examination, an outsider discovering myself through a process of accelerated learning. In the Philippines I taught at the Ateneo de Manila, the Jesuit university that runs right up from elementary school through the graduate faculties. In Iran, as a parent, I was trying to make intelligent decisions about schooling for Vanni in an unfamiliar environment, as well as teaching at two institutions and working on the planning of two others. Vanni as a child used to believe everyone had a school: Mommy's school, Daddy's school, Vanni's school, but mine kept getting shifted. At one point, the government decided to build a university in Hamadan, emphasizing local crafts and industries, teacher training, and primary health care, a regional university that would not alienate its students from their traditions—but the plan became mixed up with a project to have a university conducted in each of several European languages, so it was decided grotesquely that this local learning should be transmitted in French.

Because I am one of those people who felt at home in school and have gone on hanging around schools all my life, I keep catching myself drifting into an insidious equation of learning with education and, more narrowly still, with schooling. Setting out to talk about learning, which pervades all of life, I find myself talking about school, from which most people are happy to be liberated. Yet school casts a shadow on all subsequent learning. Trying to understand learning by studying schooling is rather like trying to understand sexuality by studying bordellos. Certainly schooling is part of the spectrum of learning in human lives, but it is not the model for all learning, only one of many byways. Learning and teaching are both fundamental for human adaptation, but not all human societies segregate them from the flow of life into institutional boxes.

Once in the Philippines I was invited to give a commencement address at an institution on the southern island of Mindanao, Notre Dame de Jolo. This was for me a curious convergence, for although the faculty were mainly Catholic priests, a majority of the student body were Muslims (called Moros in the Philippines, echoing Spanish attitudes toward the Moors). The priest who arranged the invitation hoped I would bring from Manila an association with higher education elsewhere in the Philippines and at the same time evoke the wider Islamic world. Reaching into the past, I was able to open with a few words in Arabic, recognizable to the students but not intelligible. While I was there in Jolo, I met an Egyptian, trained at al-Azhar University in Cairo, the scholastic center of Islam, sent to elevate Islamic knowledge and practice at that remote frontier. "They are like animals," he said. "They are so ignorant they hardly count as Muslims at all." There have been in human history many forms of racism, many forms of imperialism, and many forms of paternalism. Since that time, propelled by oil revenues, outreach to Muslim communities remote from Islamic scholarship has increased steadily, whether in the former Soviet Union or in the United States. No doubt increasing sophistication has led to increasing tact. The same kinds of views, with varying degrees of paternalism, were expressed by Spanish friars and by secular American administrators.

Daniel Schirmer quotes Fred Atkinson, the first general superintendent of education in the U.S. administration of the Philippines: "The Filipino people, taken as a body, are children, and childlike, do not know what is best for them."

Subject peoples are often "treated like children," in the worst sense; so, alas, are children. School is the effort to inculcate in the young, whether overtly or covertly, arrogantly or persuasively, something they could not or would not learn in their home environment, often something that alienates them from the home environment at the same time that it gives them access to a wider or richer world. For many children, learning is leaving home, perhaps never to return. On reservations, Native American children used to be separated from their parents and forced to live in boarding schools where they were forbidden to speak their mother tongues. Yet in more benign forms, the contrast between home and school is illuminating and offers an open door to a world that is wider but not necessarily separate.

Learning is the fundamental pattern of human adaptation, but mostly it occurs before or after or in the interstices of schooling. Preoccupied with schooling, most research on human learning is focused on learning that depends on teaching or is completed in a specified context rather than on the learning that takes place spontaneously because it fits directly into life.

There is another literature about learning based on experiments with laboratory pigeons and rats. This applies across species, separated from the shape of lives, and for a long time had little to say about becoming a viable pigeon or a successful rat or an inquiring human being. My father told a story of a psychologist who was asked whether, since rats are essentially nocturnal, he had ever tried running his experiments at night. "No way," he said. "They bite." "You see," Gregory used to say, "all that theory is based on the learning curves of sleepy rats." It is not that it might be possible to work out a percentage difference between the learning of sleepy and alert rats and in that way to correct the faulty learning curves. The sleepy rats were groping their way through a task that alert rats simply reject.

Gregory had extreme distaste for experimental psychology as he had encountered it, and although the field has changed somewhat, assumptions decades old still linger on in textbooks and the memories of practitioners. Another story he told was of the rat runner who decided that, since rats do not naturally live in mazes, he would try maze-learning experiments with a ferret, for ferrets live by searching for their prey in the complex interlocking tunnels in rabbit warrens. According to the story, the ferret went through the maze systematically, going down every blind alley until reaching the reward chamber, where he devoured the haunch of rabbit. The next day, returned to the maze, he again went down every blind alley but ignored the tunnel leading to the reward chamber. As Gregory said, "He'd eaten that rabbit." Perhaps the ferret had learned the complex maze perfectly the first time through but interpreted it through his knowledge of rabbits: the chamber whose occupant was recently removed would not yet be reoccupied, but any other chamber might have been only temporarily vacant, and the ferret might find the owner at home the next day.

The ferret was engaged not in an abstract learning task but in one that was intimately related to its pattern of adaptation. This is a kind of learning we know less about, learning that evokes the very being of the learner. In all the learning that involves the introduction of some alien skill, adaptive responses—seeking rewards or avoiding punishment—do play a part, but the learning itself does not match any innate adaptive pattern. No innate readiness welcomes it.

Much of traditional schooling is concerned with making children devote themselves to studies that make no sense in the context of their lives. Sleepiness is approximated by apathy, coercion, punitive levels of boredom. Research studies on human learning used to be done on college sophomores required to do tasks in the context of the classroom—the equivalent of sleepy rats. Nowadays it is more common to pay research subjects, using a carrot instead of a stick to involve them in tasks with no intrinsic rewards, and the same habit is spreading in anthropological fieldwork. Yet for a species like ours, whose survival depends upon learning, it must be intrinsically rewarding, like sex. It may be that the whole process of education prepares children for the self-alienation of civilized adulthood by turning them into permanently sleepy rats, too docile to bite.

Virtually all the learning that precedes schooling—walking, talking, bye-bye and peekaboo, the intricate rhythms of life within a household—is learning as homecoming. It proceeds at dazzling speed compared with school learning, yet it is underestimated nearly everywhere. Infants have visible states of intense alertness from their earliest weeks, and as they mature they continue to be engrossed in learning, as if they were aware of what they needed to know and how to discover it, with an unfolding promise of participation ahead.

Many people have seen photographs or read descriptions of the ethologist Konrad Lorenz followed by a line of ducklings convinced that he was their mother. Ducklings are mobile almost immediately after birth, able to wander away from nests set on the ground and vulnerable to predators. Their survival depends on learning to follow and obey a parent within a very limited time after hatching, so they are born knowing what kind of creature to look for (approximate height, waddle—Lorenz had learned to do what we aptly call a "duck walk," walking in a squat) and how to listen for a quacking sound already heard dimly within the egg. Since there is no way that the exact image of a particular parent could be supplied genetically, ducklings emerge with the analog of directions for when and how to obtain information: "when you come to the big square, look for signs." "I don't know which turn to tell you, but you'll know it because all the traffic is going that way." "You'll know it when you get there." Of course. When a particular kind of learning, like the ferret's learning of a new maze, is anticipated in the genome, new learning feels like something known forever.

We have such experiences not only in infancy, when the first moment of recognition may be lost from conscious memory, but in youth and adulthood. Learning about sexuality with a lasting vividness of delight; learning to hold and nurse an infant. There are sports where within the needed complex of skills particular components are immediately recognizable in their complete rightness, like the impact of a tennis ball on the "sweet spot" of a racket wielded just so.

Love at first sight has the same quality. Long ago I fell in love with a man who happened to stand beside me for a few seconds at the corner of Broadway and Quincy Street in Cambridge, waiting to cross; he must have matched some readiness of mine, forever unexplored. Blond, tall, thin; the image has faded with time, but for years it remained photographically impressed on memory. In such experiences, an initial, instantaneous grasp is overlaid with more gradual learning unless it is isolated or repressed.

The preservation of the image of a newborn is surely akin to imprinting, for human mothers, whose infants are not mobile, must learn to recognize them as part of the broader learning process referred to now as bonding. Usually they are lucky enough to have time and the overlaid impressions of all the senses, growing into a complex blend of love and knowledge, while the first image blurs. For years I recalled perfectly the image of my firstborn seen for only a few minutes in the delivery room in Manila, dead a few hours later. Whatever innate preparation human beings may have to be parents is probably a readiness to learn, to enter a new and strange relationship and move quickly to the certainty, This is where I belong, for this I was created. The same intense sense of homecoming often accompanies religious experience. Going back to the beginning to "know the place for the first time" must also be learning as coming home: "Yea, the sparrow hath found an house, and the swallow a nest for herself, where she may lay her young, even thine altars, O LORD of hosts" (Psalm 84:3).

It is curious that the experience of homecoming in the intuition of the sacred is then so often removed from ordinary life, segregated like much of learning into institutional frameworks that are anything but homelike. It is common to deal with moments of vision by setting them apart from the rest of experience, protecting them behind a conventional veil, whether a physical veil or a veil of ignorance or secrecy. Traditionally, the sacred has been surrounded by anxiety as well as delight. Heads must be covered or uncovered, shoes put on or taken off, eyes averted and voices lowered. Often menstruating women are regarded as too unclean to touch sacred books, enter sanctified precincts, or even pray. You can find this kind of protection of the sacred as far apart as New Guinea and the laws of the Old Testament. Jacob awoke from his vision and said, "'Surely the LORD is in this place; and I knew it not.' And he was afraid and said, 'How dreadful is this place! this is none other but the house of God, and this is the gate of heaven'" (Genesis 28:16–17). Even at its inception, awe is half horror and only half delight.

It may be that in gradually freeing ourselves from one of the traditional markers of the sacred, the recurrent tendency to wall it off and protect it at any cost, we risk losing access to such experiences, exposing them to mockery or reduction-ism or denial. But if we believe that such experiences come naturally and are basic to human beings, we may also be opening doors to the recognition of the sacred in ordinary life and in the world around us and taking back a native right. As the sacred becomes veiled in secrecy and priestcraft, sacred institutions develop that protect authority, often enforced by ignorance, and fear of the supernatural replaces the wonder of the natural. The segregation of the sacred is probably more ancient than other cultural segrega-tions of experience, for it occurs in societies with only the simplest division of labor, long before the invention of schools. Either might be a good place to start in reintegra-tion. Esoteric knowledge—knowledge that is not shared—is one of the sources of power over others.

The quality of recognition in any experience suggests a meeting of something already present within with something in the environment. We often think of the innate as a stan-dardized minimum, but the inborn and unfolding readiness to learn opens the doors to diversity of every kind: the capac-ity to grow in love for this particular man or woman, to frame experience in this language, to care for this unique and unpredictable infant. The same quality of necessity and recognition attends the poet seeking the right phrase, the painter seeking the perfect form or conjunction of colors. Artists recognize and fall in love with their own work at the point where it must be left alone. We have even made the sense of necessity a form of proof, although experience shows that what is self-evident to one mind may not be to others.

The safest and richest journeys through adolescence are those of children who discover some area of skill that becomes their very own, focusing energies and demanding for at least part of the day a honed and delicious alertness. Building model planes, ballet dancing, riding, computer hacking, basketball playing, working on a novel in secret, any of these, whether or not it promises a way of making a living later in life, can become a standard for feeling fully alive. A tool—a chisel, a guitar, or in my day a slide rule— taken up and recognized as a part of the self, can become the organizer of attention and commitment. Such discoveries, taking place outside of school, may be labeled antisocial, and children who wither in school may blossom in the acquisition of street wisdom and be punished for it. Commitment can be costly, setting children at odds with educational systems.

Because schools insist on a set range of subject matters, even those children who have fallen in love with chemistry are required to study literature and vice versa. In a society going through rapid change, a diversity of subject matter is all to the good, but it is one of the reasons why schools are at odds with the paths of learning as coming home. Colleges sometimes become so preoccupied with "well-roundedness" that they discriminate against the happy few who have, in Hopkins's words, "found the dominant of [their] range and state." We are not skilled at offering students pathways through their preoccupations to a broader perspective, as care for one child can grow into concern for all children.

The minor tragedies of lost delight in learning echo the tales of star-crossed lovers or religious martyrs. Edna Millay wrote, "Euclid alone has looked on Beauty bare," but we can only hope Euclid would have been captured by the beauty of geometry if he had encountered it in school. Most children are not; most school systems do not expect them to be. Every child who learns to walk is enraptured by the new skill, but few schools promote the same experience.

It is not that we do not value learning that comes as recognition, but that we have despaired of making it the paradigm of all learning. We mention it in shadow form when we warn that even a single dose of some drug may be addictive, may offer a sense of rightness that is forever compelling. We do not expect most children to cleave to geometry, or to the final couplet of a sonnet, as to a revelation of who they are. Yet the human species has been honed through aeons of evolutionary change for readiness to learn, in small ways as well as in the dramatic ways I have been speaking of. Each new recognition of pattern, each appropriated skill, could offer a moment of homecoming, building toward an understanding and a capacity to participate in a complex social and biological world. It is in this sense that the model of learning as coming home can inform schooling.

Most of the learning of a lifetime, including much that is learned in school, never shows up in a curriculum. When school begins much of this invisible learning is negative: the inadequacy of parents as sources, the irrelevance of play, the unacceptability of imagination. School teaches the contextualization of learning and the importance of keeping different areas of life separate: home from the workplace, Sundays from weekdays, and work from play.

The knowledge that children bring with them into school has not been learned in an orderly progression. It can be codified and systematized (and sometimes is by linguists or anthropologists), but it is mainly passed on in contexts where it is presented not in explicit linear sequences but through spirals of partly apprehended repetition. Learning to speak implies grammatical rules and category systems, ways of mapping and classifying the world. Children's rhymes and stories contain metaphorical statements about the structures of the real and the social worlds, often coding vast stores of information. Childhood has its geography and natural history, its ethics and metaphysics, not without pain and effort, but often without alienation.

San children grow up with an intimate knowledge of their environment, a complex grammar and mythology. Ties between persons are coded in three kinds of overlaid kinship and naming systems that take up several pages of diagrams in an ethnography. San children never see the diagrams but instead see living patterns of gift giving and mutual aid, gradually sorted out in the course of childhood.

The San have no indigenous tradition of schooling, no

professional teachers, but like every human community they do teach. We are the animal that relies most on learning in our adaption, and even more distinctively the animal that relies most on teaching to evoke a portion of that learning. Just as the long human infancy requires reliable adult care, so the learning of survival skills requires reliable adult teaching: human biology depends on love. A San father takes his son out on the veldt with a spear to learn to track wild animals, just as an American father takes his son to the park to learn to hit a baseball. A village mother in Iran may give a warning or a demonstration before a daughter is allowed to use a loom or a sewing machine, wool or butter, knives or fire. Often what is taught would not be learned if it were not embedded in a relationship, for it may have no obvious relevance: a child's hands may be moved through gestures of ceremonial, the sign of the cross or the beginning forms of dance. A parent may teach a child the words of a prayer, presenting it line by line for memorization, often enough in an unknown language. "In the name of Allah, the merciful and the compassionate," "Hear, O Israel," "Our Father who art . . . "

Much that looks specific is really general instruction in relationship. In Western societies, we overestimate the importance of odds and ends of explicit teaching, without noticing what is learned implicitly. When we teach "Don't say 'it's me,' say 'it's I'" or "Say 'thank you'" or "Don't scratch in public," we are using relatively trivial explicit teaching as part of the process of imparting informal knowledge of a highly abstract kind about correctness, public and private spaces, and the nature of authority. Educated parents put considerable effort into correcting certain "classic" errors of grammar ("It's me") while blithely ignoring complex syntactic processes that children master without ever having them explained. Similarly, parents spend considerable time telling children to say "please" and "thank you" without instructing them in the more subtle gestural courtesies and alternate forms they will eventually master. We tell our children to say, "Please pass the butter," and almost unnoticed they learn to say, "Could I have the butter?" in a tone that makes it accepted as equivalent. Clearly the lesson in courtesy is a vehicle for another less explicit and more profound lesson, like Parvaneh simultaneously instructing Shahnaz to be friendly and to withdraw from strangers. The informal learning, unverbalized and unquestioned, takes precedence over explicit teaching unless uprooted in drastic ways.

The same is true on matters of values. We instruct our children not to hit, not to make another child cry, and to "be nice to the little girl," but by example and other subtler clues we also instruct them that in some cases they should hit back and that they should be nicer to some people than to others. Subtle lessons about how social structures really work are passed on to children before they go to school, often before they are exposed to more presentable but contradictory verbalized values, which may then prove extremely difficult to teach. Sometimes when I paused to chat in the Philippines, a mother would say, "This is our fair child and this dark one is the ugly one," and I would be filled with white guilt and play with the darker child, knowing that there was little I could do to modify an often repeated lesson that would haunt both children for life. If African American children are told, with yanks and impatience, that they have "bad" hair, they may learn a much more general lesson of badness. Joan, with her educational toys, form boards, and color books, was teaching Becky something, although not yet the lessons visualized by educators.

Discovering the connections and regularities within knowledge you already have is another kind of homecoming, a recognition that feels like a glorious game or a profound validation. When I started describing cultural patterns that were creating conflict between Iranians and Americans, one U.S.-educated Iranian said with pleasure that my analyses made his own informally learned traditions seem "reasonable," for the first time. If teachers were to approach their classes with an appreciation of how much their pupils already knew, helping to bring the structure of that informal knowledge into consciousness, students would have the feeling of being on familiar ground, already knowing much about how to know, how knowledge is organized and integrated. This might be one way for schooling to assume the flavor of learning as homecoming: learning to learn, knowing what you know, cognition recognized, knowledge acknowledged.

When schooling conflicts with previous learning on specifics, more general patterns may be disrupted and the sense of how knowledge is put together may be unraveled. So often, schooling depends on the idea "Take care of the pence for the pounds will take care of themselves," but the pounds are the fundamentals. An American child who has been told to drink her orange juice when she has a cold has learned exactly the same truth about the process of learning as an Iranian child who has been told never to drink orange juice when he is sick: that appropriate behavior must allow for all sorts of invisible relations of cause and effect, taken on trust. Better theories of nutrition are not a fair trade for impaired trust.

Eating carrots helps you to see in the dark. Garlic repels vampires. Cholesterol causes heart attacks. We all accept a vast number of such beliefs, and simply attacking those that have not been empirically validated creates confusion. The message "you are ignorant" is an attack on all the learning gained up to that point, not just on particular errors. It is more important to learn ways of grasping and organizing and testing such propositions, in the context of an affirmation of the process of learning. We may yet get a different version of the cholesterol story.

It has been said that the most important intellectual achievement of any human life is learning a first language, yet, except for brain damage, this is something we all have in common. We all enter school speaking a first language. In school we find out its name. A child who has learned to speak a nonstandard form has learned as much about *how* to learn a language as the child who has learned a standard one; that learning to learn has to be conserved. When a child enters school, even where the language of instruction is very close to the language of the home, he or she is still at risk when teachers spend their time teaching *correct* forms instead of celebrating the fact that every child already speaks some language pretty well. The structure of school emphasizes what you don't know.

A great deal could be gained from the traditional first language classroom by making systems learned without being verbalized explicit. If this could be done without devaluing the earlier, unverbalized learning, it could make available an additional layer of learning to learn. For example, English speakers go through their entire education without ever becoming explicitly aware of the rules they use, regularly and accurately, for forming plurals in spoken English. Children know this stuff when they arrive at school; all they need to learn is how to spell it and how to handle a few special cases. But discovering the rules offers a kind of self-knowledge, the discovery that one is cleverer than one knew, in ways one had never noticed.* Sometimes teachers are unaware of their own unverbalized knowledge and take it for granted as a foundation, failing repeatedly in the attempt to teach pupils from other backgrounds in whom that knowledge is absent or different. The fact of patterning is far more important than knowing which pattern is in fashion in a particular period.

English teachers, even while offering the standard alternatives, could honor the elegant patterned quality of many "mistakes" (such as the use of *like* in reporting dialogue: "He's like, 'What can I do?' and I'm like, What can I tell him? And he's, 'Maybe I'll go home.'"). Black English uses its own set of variations with equal regularity. Affirming patterns already learned would mean a profound modification of the teacher-student relationship: skills achieved could be built upon or varied rather than replaced and students could be treated as expert sources on their own experience. Conflicts between home learning and school learning could be

---

*Collect examples to show that the plural suffix written *-s* or *-es* has three pronunciations; which occurs where? To confirm that the choice of pronunciation depends on meaning as well as anatomy, compare the sounds of *knees* and *niece* or *whores* and *horse*. Then find the same pattern of variation in the regular suffix on verbs indicating the past, written *-d* or *-ed*. How elegant it is that the patterns are so similar.

replaced by comparisons of alternative patterns instead of a dissonant jangle. Schooling could offer the chance to choose behavior that will be adaptive, rather than forcing it.

The rules for how different kinds of knowledge fit together, which allow for the transfer of knowledge from one situation to another and for what linguists refer to as the generation of novel performances from underlying competences, are especially likely to remain unstated. Skills in seeking out and judging information that are explicitly taught are the tip of an iceberg whose base is formed when children learn to distinguish between fiction and news on television, to formulate the thousands of questions toddlers ask, to choose an adult likely to give intelligible answers, and to understand why some people are annoyed by questions and others pleased.

Everywhere in the world, the contexts of learning change with maturation, switching from play to courtship to ritual. Cultures have mechanisms to accelerate learning at key points in the life cycle that build on the ancient link between learning and altered states of consciousness, like those that often form part of initiation rites. When adult participation in a society requires unlearning something already learned, the pedagogy may be draconian, yet often children accept it as a necessary transition to adult identities, part of becoming themselves. Without physical mutilation and fasting, we too maintain solemnity and unpleasantness in schooling, and insist on undoing earlier identities and confidences. Teaching children that there is a correct time and place for learning, we also teach them to *stop* learning when they manage to escape from school, or to keep what has been learned specialized to one context and quite inaccessible for use in others, like tourists who become tongue-tied in Paris after years of high school French.

The polarity between initiation and alienation recurs in system after system; so does the polarity between persuasion and coercion. Often in missionary situations education is focused on matters that parents agree in wanting their children to master but comes blended with material that would alienate them. There is a profound tension between the idea of learning as coming home, carrying with it a steadily widened definition of home, and the idea of learning as leaving home. Jewish children exposed to Christmas celebrations at school, African American children recruited to privileged schools from the ghetto and then isolated, Muslim children in the Philippines, or Armenian children in Iran have no ready escape from ambivalence. There is a thread of betrayal in schooling of every kind.

I have never been anywhere that education was more hotly pursued than in the Philippines when we were there. The American colonial administration put its emphasis on economic and educational improvement and public health. Already in 1901, a shipload of six hundred American teachers had fanned out across the country to establish a free system of public education. They were called the Thomasites after the name of their ship and are praised and excoriated by turns, for they were both a boon and a curse, muting the sentiments for resistance and independence. The Thomasites were like an early version of the Peace Corps, opening schools in rural areas and beginning a wave of literacy affecting the whole country. In the Philippines literacy and knowledge of an outside language are still extraordinarily high for a third world country.

The Republic of the Philippines has been independent since 1946, but it relies on English as a lingua franca alongside Pilipino. For children beginning school in their home dialects outside the Tagalog area, then, literacy involves at least three languages. As in India, there is a distinctive form of English filtered through generations of locally educated teachers: fine in context, it requires modification for export, suggesting the need to learn still another form. Here is a sequence of language learning that in principle follows the pathway from identity to adaptation. When languages are separated by context, children can master more than this; when they are muddled or disaffirmed, the whole process can be inhibited. Some African Americans can move skillfully up and down the scale of variations, from a deep southern black dialect virtually unintelligible to white northerners to BBC English, playing with the music of the differences. Others feel trapped in a pattern of speech that labels them as ignorant, wounded in their sense of who they are and ill equipped to adapt to others. Standard English, so useful for many purposes, has come to seem an imposition to many for whom it could offer a useful second string to their bow.

Persuasion or coercion? Proprietary higher education became one of the most profitable businesses in Manila, sometimes superb, sometimes a shoddy and exploitive product. In spite of unemployment in many fields and mismatches between the supply of graduates and the need, everyone seemed to believe that education was the key to advancement. When the Marcos government was overturned, there were students in the vanguard.

The experience of the Jews has been very different, because for them schooling and study have been central to identity for millennia. It is among Orthodox Jews that the joys of learning are most vividly affirmed. In Israel little boys in black with skullcaps and side curls, growing up in the Orthodox enclaves that most nearly replicate the ethos of the Eastern European shtetl, can be distinguished from the children of secular families by posture and coloring, for these are children who do not play in the sun. The poet Bialik described the house of study as a prison, as something rotten and emasculating, yet he wrote, "who are you, adamant, who are you flint, to a Hebrew boy occupied with Torah?" Whereas the popularity of study in the Philippines is largely instrumental and education and certification are pathways to prosperity, in the Jewish tradition, scholarship has been an end in itself: the Torah was not to be used as "a spade with which to dig." Learning Torah is pure delight. Wealthy men coveted scholars, the true aristocrats of the community, as sons-in-law and were ready to support them in a lifetime of study. In Israel today a reinvigoration of traditions of Torah study is a central theme of the resurgence of Orthodoxy. Learning Hebrew and Old Testament were, even for me as a non-Jew, formative intellectual experiences of self-discovery, made possible because when Hebrew was revived from scholarly use to become the living language of Israel, this meant the creation of a community that welcomed and supported the language learning of adults.

Learning as a tool. Learning as an act of worship. Learning as a betrayal. Learning as play. Learning as servitude.

Learning as a way of life. Education in the Philippines both empowers and disempowers. It is both a distribution of wealth and an investment, portable after political turmoil. Where land reform fails to put the basis of prosperity into the hands of peasants, schooling can still do so, and education, unlike landownership, can be shared by all. Knowledge can represent both domination and humility, courtship and combat. Education creates a malleable and skilled workforce, but it also perpetuates elites and creates revolutionaries. It can create xenophobia or cosmopolitanism. In interviews given by the shah in the late sixties, it is possible to detect a tenuous and dawning awareness that modernizing education was the only way for Iran to go but that the process could eventually end the monarchy. After the Islamic revolution, schools and universities were closed for months and even years so the education system could be reconstructed to match the ideology of the new government. They knew all too well that education is not just about literacy and numeracy, that it has always been contested ground, the stuff of power and identity.

I never found a fully satisfactory answer in Iran to the conundrum of schooling for Vanni, but then, I have never found a satisfactory answer to what I am doing in my own teaching. In Iran, the local schools seemed to me so preoccupied with issues of authority and correctness that they suffocated creativity. Secular American schools were separated from the society around them, while mission schools had too much hidden agenda. Eventually we turned to schools founded by American or English wives of Iranians, hoping that in at least a few of those marriages there was mutual respect, continuing learning, and an effort to find and make a home, a home which could provide a model for the schooling of bicultural children. Even when both parents come from the same background, a successful marriage is a continuing learning experience, constantly involving communication across difference so adaptation does not threaten identity. In successful bicultural marriages, cultural differences enrich the process.

I still teach for a portion of every year, puzzled by the ambiguities of the enterprise. A professor is supposed to be authoritative and well prepared, so it is hard to resist offering answers without questions and conveying the message that the world is divided between those who know and those who do not. My own greatest resource as a teacher is the learned willingness to wing it in public, knowing that I will be faced with unexpected questions, some of which I cannot answer. This is the challenge—improvising, learning on the job—that my students will confront all their lives. Oddly, I find myself trying to convey two contrasting ideas. On the one hand, I try to teach students to benefit from difference instead of being put off by it. On the other hand, I find myself discouraging the notion that learning depends on that specific difference we call authority.

Today there is a wealth of new thinking about schooling, yet it is fashionable in America to say that schools are failing and there is a groundswell of anger against educators of all kinds. This is not in the main because they are not doing their job—it is because we have no adequate understanding of what that job is in the kind of society we are becoming. We think the issue is the transmission of specifics, the meeting of specified goals, but these are illusory and children are wise enough to know it. It is a mistake to try to reform the educational system without revising our sense of ourselves as learning beings, following a path from birth to death that is longer and more unpredictable than ever before. Only when that is done will we be in a position to reconstruct educational systems where teachers model learning rather than authority, so that schooling will fit in and perform its limited task within the larger framework of learning before and after and alongside. The avalanche of changes taking place around the world, the changes we should be facing at home, all come as reminders that of all the skills learned in school the most important is the skill to learn over a lifetime those things that no one, including the teachers, yet understands.

It may be that withholding commitment and retaining skepticism even in the classroom is the wisest course, for we cannot tell our children with conviction that the civilization we know will always be right or true. We know it must change. "When you come to the big square, look for signs." For that looking, we can provide models for multiple kinds of attention, not attention paid like a tribute to an enforced lesson, but attention claimed and honed as a right of entry and a rite of initiation. You will always be acting under uncertainty. You will know the future when you get there. Only so can you make it your home.

There is another sense in which learning can be coming home, for the process of learning turns a strange context into a familiar one, and finally into a habitation of mind and heart. The world we live in is the one we are able to perceive; it becomes gradually more intelligible and more accessible with the building up of coherent mental models. Learning to know a community or a landscape is a homecoming. Creating a vision of that community or landscape is homemaking.

14

The Seemly and
the Comely

LIKE MANY CITIES in the third world, Manila has huge areas of slum and squatter settlement, of which the most famous is Barrio Magsaysay in Tondo. Back in 1967, when my husband and I were living in Manila, I considered doing fieldwork there and visited several times with Filipino social scientists, but in the end research in the Marikina Valley seemed more workable. In those brief forays, however, visiting both squatter shacks and housing projects in Tondo, attentive to the dissonances imposed by extreme poverty and the need to make do with whatever is available, I found myself noticing harmony as well and seeing patterns in a different way.

In remote rural villages, Filipinos traditionally live in small cubical dwellings raised on stilts and built of nipa palm and bamboo, with peaked roofs that drain off the heavy tropical rain. In the provinces you sometimes see elegant old wooden houses, built by the aristocrats of previous generations, preserving some of the same proportions on a larger scale. The windows are not glass but dozens of translucent white capiz shells mullioned in square frames.

In Barrio Magsaysay, houses are made of shipping crates,

scavenged billboards, flattened oilcans, and odds and ends of wood and metal. Walking between the haphazard rows of ramshackle dwellings, with their patchwork textures, your first impression is one of chaos and probable violence. Yet, just as you can sometimes recognize an adaptive pattern best when it is transferred from its original setting—watching a dog or a cat turning in circles a few times before lying down on a carpet where there are no ferns or grasses to flatten— the makeshift betrays an underlying pattern. It suddenly occurred to me that again and again, in the patchwork of squatter shacks, I was seeing not quite perfect squares. Window openings were square, walls were square, sections of fencing around tiny yards were square, doors were frequently divided into two near squares. The effect of this was that the hodgepodge of slum dwellings presented a spontaneous aesthetic order, one that I noticed more readily because it was strange to me, for the architectural aesthetics of the West are based not on squares but on rectangles. Perhaps my liking for patchwork quilts today comes from this recognition of an aesthetic order worked into the improvised housing of the very poor.

On the outside of many of the shacks there were skillfully knotted and twisted ornaments made from the palms of Holy Week, like the woven ornament I have now on my studio wall. Rows of colorful croton plants and bougainvillea were often lined up in big, rusty oilcans, and the bushes around the houses were decked with emptied eggshells, like glowing white buds. Shacks were crowded together, blocking access for firefighting, unsupplied with electricity and running water. Yet working with very limited materials, the people of the district, migrants from the countryside, were constructing a world that was, what? Comely. Seemly.

Once I began to look in that way, it was possible to see the lingering preference for near squares in other settings where the indigenous aesthetic was almost overwhelmed by imported styles, the vertical rectangles of office blocks and the horizontal rectangles of wealthy villas. Barkev and I bought a painting by a young painter named Lamberto Hechanova, an abstraction in red and black of a vaguely angelic figure on a square canvas. When I returned some

time later to see more of his art, I found a large work made
by cutting and hammering tin cans and nailing them to a
near-square piece of plywood, all painted gold, a bas-relief
mosaic of shacks with the sun rising above them, which
Lamberto, a child of Tondo, had titled *Exultant Slum*—exultant in the gold of sunrise, or perhaps exultant in the survival
of an aesthetic sense of rightness that made houses, however
they were constructed, into homes. The painting hangs today
on our dining-room wall as a glowing reminder of the way
human beings create pattern and order even under the most
stressful circumstances. Friends tell me that they have
learned to look at the favelas in Rio de Janeiro and Brasília
as well to see emerging physical and social order as gardens
and ornamentation appear.

The people of Barrio Magsaysay would not have decorated their houses if they did not feel a certain pride and
pleasure in them, but they would certainly prefer to live otherwise and at one level would concur with outsiders who see
the squatter areas as the expression of social inequity and a
seedbed for crime. Squalid. A national shame. We can celebrate the fact that human beings weave pattern and order
under terrible circumstances without romanticizing the
social systems that put them there. But unless the patterns
are recognized, reforms destroy more than they improve.
Worse, the effort to help by imposing alien patterns has hidden in it a profound pessimism about human beings. In
addition to failure to see emergent order, it is common to see
the very behaviors that allow people to adapt and reach for
dignity as symptoms of breakdown. Whether in social life or
in ecology, the impulse to improve without first understanding is dangerous.

We are still haunted by the compelling falsehood of
Hobbes's description of the primitive as the war of all against
all with "no arts; no letters; no society; and which is worst of
all, continual fear and danger of violent death; and the life of
man, solitary, poor, nasty, brutish, and short." Many who
would no doubt reject the myths of Genesis and disavow a
belief in original sin accept Hobbes's superstition about the
state of nature: that the weaving of order is not intrinsic to
human beings and so must be imposed. A narrow system of

education based on a single set of cultural preferences is then defended as all that stands between us and chaos. It is as if Westerners had achieved the recognition of other members of the species as more or less human without appreciating the web of meaning this everywhere implies. That kind of limited recognition is a license to tinker and disrupt human communities, as technology has disrupted natural ecosystems, also in the name of improvement.

Carol Stack, in *All Our Kin*, explored patterns of sharing and fostering in a black urban ghetto in the sixties, under conditions of poverty and vulnerability, as cash, furniture, and even food followed need from house to house, recycled garments gave the thrill of novelty to multiple wearers, and children nested where there was an adult available to care for them. Such patterns are harder to recognize than the geometries of Tondo, since they demand that an observer stay to note daily behaviors, looking and listening instead of administering questionnaires. Unfortunately, the very same adaptive order that Stack described violates the regulations of welfare workers and feeds stereotypes of illegality and fecklessness, slipping like water through the structures imposed by external pressure. Searching for intact nuclear families, the authorities disrupt networks of kin. Bemoaning the lack of a resident father, they may fail to notice the contributions made by a resident grandmother—and impose regulations that force the father to stay away.

Culture of poverty should have been a useful term for describing what Stack found in the ghetto and what I saw in Tondo, for the very word *culture* celebrates the human capacity to learn and adapt, something the rest of society should support. However, the term has lost this potentially positive connotation and come to be used to refer to anomie and alienation, to attribute inevitability or even preference to the compromises of those forced to make do with so little. The tenuous adaptations of the poor, order achieved and some pattern of meaning maintained against all odds, have themselves come under attack, and life in the ghetto has worsened. True, the patterns Stack described may not transfer well out of their environment, may hinder emigration or provoke conflict at the boundaries so that outsiders see them as criminal and walk in fear, but the culture of poverty must be understood above all as the achievement of survival.

Perhaps all those who work for change should study jujitsu, learning in their very bodies to work with the impulses and directions that are already there, rather than opposing them. In Tondo, as in many other places, massive, impersonal housing projects were built to replace squatter shacks. Inside their tiny, bleak apartments, the residents built inner shells of bamboo, and women opened little shops in the doorways, corresponding to the shops along village streets that become focal points of gossip. When I was there the authorities kept closing them down because it was against regulations to do business in public housing, instead of seeing that they had the potential for turning dangerous empty hallways into village streets. In the same way, in schools we accuse children of cheating when they work together instead of seeing that the skills of working together would be worth enhancing and supporting.

Under most circumstances, human beings persist in their evolutionary business of sustaining and creating order, doing even trivial things in a particular way that feels right to them. A sense of coherence is almost as needful as food and drink. On the streets of Tehran, merchants arrange small goods of various kinds—combs, cigarettes, and so on—on cloths. After a time I realized that these impromptu sidewalk displays reflected an aesthetic of multiplicity that could be seen in the array of items on a traditional *haft sin* table arranged for the Persian New Year, which must include seven items whose names begin with the letter *sin*, and in the arrangement of items spread in front of a bride and groom.

Finally I saw that these "arrays," as I came to call them, echoed the desired sense of bounteous hospitality in the multiple items in a meal. A festive meal in Iran is not organized around a single climactic pièce de résistance, like a roasted turkey or a beef Wellington, just as the New Year celebration is not focused on one major symbolic item, like a Christmas tree. Multiple dishes are set out, each one divided among several serving containers to be within reach, and ideally the table is still crowded with food at the end of the meal. Waste in the American tradition is reduced by recipes for recycling leftovers in new forms, but in the Iranian tradition it is reduced by the dissemination of portions to servants and neighbors.

Trying to improve people by interfering with their own preferences often makes things worse. One of the behaviors that is most easily condemned in other communities is drug use, although chemical tinkering with mental states is very nearly a human universal. Almost all human groups have found something to eat or drink, to sniff or smoke, that alters moods and even metaphysics, but these practices are less dangerous when they are regulated by tradition, suddenly more dangerous when suppressed. It is often custom, not chemistry, that determines whether a practice is harmful, and many interferences disrupt custom and leave chemistry to do its worst.

In Iran, many adults used to use opium much as Americans use alcohol, understanding that smoking opium was dangerous for some but a relatively safe social habit for most. We used to be invited out for a family meal on a Friday: great platters of delicious foods, lamb cooked with quinces, duck with walnuts and pomegranate, rice with currants or shredded orange rind. Afterwards an ornate charcoal brazier would be lit to provide the embers to vaporize crumbs of opium. Whether one chose to smoke or not, there was pleasure in the setting, Persian carpets and cushions, conversation and laughter, multiple cups of tea and varieties of sweets. Smoking too regularly was regarded as risky, but not inappropriate in the very old, those who were sick or arthritic. Others who had abandoned the courtesies of the shared brazier seemed to be the same people, often highly Westernized, who smoked urgently and frequently for intoxication, abandoning the protections of good manners, like the solitary drinker for whom alcohol is no longer social. In the same way, when beer is forbidden on college campuses, students are likely to keep bottles of hard liquor in their rooms. Alcohol has different meanings in different settings; surrounded by formality or ritual, tied to connoisseurship or mysticism, it is less toxic to body and to community. Historically, in most parts of the world, observant Jews, who use wine weekly for the ritual of the Sabbath meal, have had few alcohol-related problems.

Western civilization tends to cut items of behavior out of their matrix and to find ways of concentrating and intensifying substances: wine with food is replaced by the cheap distillation of gin, the chewing of coca leaves is replaced by the use of cocaine and then crack, opium by morphine and then by mainlined heroin. Native Americans traditionally used tobacco with respect, as a divine gift. Disapproval makes some habits more dangerous as they are forced into secrecy and pursued without social controls. Prohibition isolates users and makes illegality tempting and habitual, creating whole classes of bootleggers. Breaking rules is bad for the health of the breaker.

Courtesy, propriety, standards of order are mechanisms of survival. Theodora Kroeber has described how a tiny remnant of Yahi Indians of the Mt. Shasta area in California, nearly wiped out by Euro-American invaders, went into hiding in the forest and survived, unknown to the whites around them, for years, until the last survivor staggered out, sick and starving, into the unknown white world, where he was sheltered and adopted by anthropologists. Since it was not customary in his tribe for any to pronounce their own names, he is known to us only as Ishi, "man"—or, if you will, Adam, not the first but the "last man left awake." But clinging to survival on familiar ground even as their numbers became fewer and fewer, Ishi's people preserved their standards of decency, although many customs had to be adapted to life in hiding. There was a young woman near Ishi's age whom we believe to have been some kind of second cousin, but because, in his traditional kinship system, even when the community was reduced to half a dozen members, a cousin of that kind was classified as a "sister" and barred by the incest taboo, Ishi never married. There was no one else. The Yahi understood survival to mean survival not as biological organisms but as the carriers of certain patterns of order and meaning, intrinsic to themselves. This is a far cry from the legend of Lot's daughters, after the end of humanity as far as they knew it in the destruction of the Cities of the Plain, seducing their father in order to preserve the future.

We have come to lack faith in the resilience and creativity of human order so we lack too the willingness to recognize it where forms differ. Because we do not regard the rudiments of culture as intrinsic to humanness, we keep comatose bodies ticking in hospitals and nursing homes and many argue that the fetus is fully human from conception, before learning and before participation. We tend to believe that the incest taboo has to do with preventing genetic flaws, whereas in fact it is one way human beings everywhere have tried to conserve the trust necessary for social life. With a similar myopic reductionism, I have sometimes heard Iranians argue that the prostrations of Islamic prayer exist to provide rudimentary calisthenics. They may indeed be helpful in keeping joints limber, but the limberness of the spirit is more important. No reductionist interpretation of Jewish dietary laws as a substitute for refrigeration should be allowed to obscure the fact that they order life and preserve community.

It is perhaps because we have not learned to recognize and respect existing order in unfamiliar forms that we are frightened of social change, unwilling to support and work with the forms that peoples find for themselves. What stands between members of our species and social chaos is not reading the Bible or the works in the Western canon but the habit of patterned relationship, a habit that shows up first in the mutual adjustments of mother and infant adapting to each other's rhythms. Imperfect and incomplete for many, this mutuality is still sufficiently pervasive to provide most human beings who survive to adulthood with an experience of how separate organisms can interact in harmony. Robert Edgerton has described how even the severely retarded, unable to learn language, show a capacity for friendship, loyalty, and caring that makes one ask how to prevent schooling from destroying values rather than how to inculcate them, for social life has a model of harmony at its very inception. Without it we would not have survived all these millennia.

All too often, members of one group look at another from

the outside, misjudging patterned and adaptive behavior as lazy, ignorant, or licentious. In Iran I collected lists of Iranian behaviors that were irritating or worse to Americans—and similar lists of American behaviors offensive to Iranians—then tried to find ways to make each one intelligible by explaining where it fit into larger patterns. This was in the days before self-serve gas stations in the United States, so Americans arrived in Iran with the expectation of finding "service stations" where gas would be pumped for them, windshields cleaned, and a wide range of services available. Gas stations in Iran sold gasoline only, and the attendants were there as cashiers. If they gave personal assistance with the pumps, they expected a tip. Americans sat luxuriously in their expensive cars, expecting to be waited on, then drove away without giving the attendants anything, so the attendants dragged and tended to shortchange to get what was due them. Two cultures, each seeing only a part of the other's pattern and attributing it to character flaws, arrogance, dishonesty, laziness. "Typical," each would say of the other. "Just like them." All too many Americans responded to the small book I wrote about contretemps of this sort and how to avoid them, Why should we understand their culture? Let them learn to understand ours. But my first task was to demonstrate the very presence of order.

It is the business of professional anthropologists to winkle the patterns out of voluminous notes and months of work and to analyze them. The question for everyone, living in a world of constant contact between cultural groups, is how to become routinely sensitive to patterns, even with minimal cues, suspending judgment and looking for how they fit together.

I know only two ways to prepare others for that kind of attention. One is by offering early and often the experience of difference, always in the context of the expectation that there will be a pattern to observe. The United States presents unprecedented opportunities for learning about the range of human potential, if only we look with open eyes. The other is by offering early and often the experience of making and looking at art, which demonstrates that two people can look at the same mountain and see something different. Rudyard

Kipling once wrote a poem in which he described heaven as a sort of celestial artists' colony; I thought of it often when I was at the MacDowell Colony, where ". . . each for the joy of the working, and each, in his separate star,/Shall draw the Thing as he sees It for the God of Things as They Are!" Kipling's God sees a world we can never see, but we can sense its mystery by being open to alternative visions, including those offered by the telescope and the microscope along with those offered by painters and poets and other cultures.

Did sunflowers look to van Gogh the way they look in his paintings? Not exactly, clearly, but they apparently did look different to him, and in interpreting that difference and translating it to the canvas he constructed a greater difference. Of course we cannot see either sunflowers or paintings as he saw them, and our way of seeing is different again from the way his contemporaries saw. It is fashionable today to attribute the unique vision of someone like van Gogh to organic causes, yet van Gogh changed sunflowers forever. They keep on changing. Today we increasingly see van Gogh's paintings framed in dollar signs, and those who see them framed in yen must see something else. At the same time, having once seen a few paintings by van Gogh, we can recognize others.

The special case of the artist's idiosyncratic vision expressed in work after work provides a partial parallel to the stylistic coherence that allows one to say that a particular wooden bowl was carved in the Trobriand Islands, or that a particular rug is the work of Qashqai weavers, or to divine the cultural background of a particular street merchant or the underlying vision of home sustaining a shanty builder. In complex societies we emphasize the uniqueness of the work of the individual artist, but in traditional societies the shaping of pottery or tools and the ornamenting of the body provide bridges that outsiders can see more clearly if they look at them as art, while the members of the community see someone doing in a seemly way what has always been done, perhaps in a manner just a little more comely in the hands of this potter or wood-carver. Some enact even very traditional and homogeneous cultures with more grace than others.

These two issues converge in modern society because of

the difficulty in doing things as they have always been done in a changing setting, so that individuals are obliged to invent new ways of composing the elements of their lives. Today survival depends on a willingness to move away from familiar patterns, but we do not emphasize often enough that new patterns must satisfy ancient needs for harmony and that familiar graces also contribute to an evolving aesthetic of adaptation. The attention that looks for unfamiliar kinds of order even in behaviors that appear outrageous or bizarre may be a precondition for the capacity to generate new patterns from unfamiliar materials. Every time there is a brouhaha about censorship in the arts, like the outrage of the Iranian mullahs at the work of Salman Rushdie or of the American right at the work of Robert Mapplethorpe, I think about the double rhythm of pattern violation and pattern creation in the arts that must also characterize any society undergoing change. We live like the people of Tondo, torn out of traditional patterns and attempting to make sense of our lives by composing and creating new ones that will seem good to us with whatever materials come to hand.

Works of art that offer new ways of seeing the human body propose new ways to be human. American society has been forced within the space of a generation to accept the presence of a large gay community and to grant its members rights of privacy and self-determination, but this acceptance is grudging and graceless until we learn new ways of seeing. The work of an extraordinary generation of homosexual novelists and poets and photographers allows outsiders—straight people—briefly to see that what society was grudgingly forced to tolerate is in fact both seemly and comely, to acknowledge different kinds of grace and tenderness. The urgency of this acknowledgment is sharpened by the AIDS epidemic, which had its first rapid spread in America in the gay community, for curbing the spread of AIDS has required ever stronger patterns of mutual caring and respect both within that community and between the gay community and the straight majority. It is in this sense that experimentation with new kinds of beauty has survival value.

The move toward racial justice in the United States had to include an understanding by both blacks and whites of the beauties of black men and women, the richness of dark skin, the grace of bone and muscle, the coherence of nappy hair. Here, too, the experience of newness may be essential to developing a new aesthetic. I have seen racial differences and racism in a new way since I lived in the Philippines, because I discovered there the variability of my own visceral responses. After living for months in a village with dark, graceful, small-framed people, I returned abruptly to a gathering of white Americans who suddenly seemed to me bloated and unhealthy, like beached white whales decaying in the heat ... later, I was confined at home by illness for two weeks and felt the shock and oppression of alien difference the first time I went out. If nothing else, these moments of self-recognition taught me that the reactions that underlie racism (like ethnocentrism and homophobia) are a human heritage not limited to the bad guys, and that such reactions can be learned and unlearned and mined for insight. Even those artists who concentrate on shocking offer an opportunity to learn, but if that is all they are doing their work will seem trivial soon enough.

Those who try to limit artists to confirming existing ways of seeing may also limit their own capacity to learn. The rejection of the unfamiliar in the arts is often a blend of imposed tests of reality and propriety. If observers say, from their limited point of view, It isn't really like that, the work is false, slums are terrible places where life is nasty, brutish, and short ... they are introducing something intended as a reality test. Art that is reality-tested in this way is not allowed to reveal new realities or to foster in the observer the capacity to discover and construct new realities elsewhere. We are left vulnerable to a second order of reality testing, challenging the flexibility and adaptability of the society, the reality testing of evolution.

If an observer labels a work of art as blasphemous or obscene, he is imposing a propriety test, in which some possible perception is to be unmentioned, not unknowable but unspeakable. When art is propriety-tested we lose the opportunity to look for another kind of propriety, the seemliness in the artist's own vision, new kinds of harmony or proportion that seemed outrageous on first hearing. When any society protects itself by making large areas of experience unmentionable, using ignorance to control behavior, it takes on the risks of action in ignorance, like adolescents discovering sex with no awareness of its dangers or potential. Behavior is isolated from the effort to shape it into forms that can be shared.

My mother, who came from Pennsylvania, used to say that the strongest form of disagreement the Quakers she grew up with would allow themselves was to say, "Friend, it never would have occurred to me quite that way." Whatever the consensus finally forged by the group, there is a discipline implied in that phrase "it never would have occurred to me," the suggestion that perhaps I could learn to let it occur to me. Perhaps even then I would reject it, but I would be enriched by having tasted a different view of the world.

Walking through foreign streets as a tourist is refreshing, but once in a while something brings home to the visitor the differences in vision so sharply that the world is re-created. Japan has been for many Westerners a place where the sense of cultural difference is immediate, with little temptation to mask it with the illusion of superiority. In Bunraku puppetry, the puppeteer is fully visible on the stage. This cannot be because the Japanese audience is unable to see him but must offer a clue to a different map of the relationship between "reality" and "illusion." In some Japanese gardens a guide or companion will explain that the garden demonstrates the Japanese love for nature, even as you see cables that pull branches into their appointed directions, the endless care taken to exclude every weed and keep the gravel paths just so: a different map of the relationship between "nature" and "artifice." As we enter the era of virtual realities, it is well to remember that these will reflect not only the desires but also the assumptions of their creators. Yet all peoples—and all individuals—learn to understand the world by learning to invent it, and it is in the invented world that we must survive. Our best chance of survival lies in seeing and inventing that world as beautiful, inventing it with the precision of wonder.

15

Reflected Visions

WAITING IN THE TRANSIT LOUNGE of the Athens airport, on my way back to Tehran from a trip to America, I saw, out of the corner of my eye, a group of Tibetan Buddhist monks: tall, golden men with shaved heads and biceps bulging in deep-wine-colored robes. There was another man, not a monk, with them, a Caucasian, and he was interpreting for them at the service desk. When I drifted closer to listen in, I found that they were en route to Tehran also, in fact on my flight, and I decided to pick them up.

When China invaded Tibet and broke up the great monastic communities, many of the monks fled to India. In the years that followed, the monks arrived at an understanding of their disaster that gave it meaning. Prior to that time, Tibetan Buddhism had been limited to Tibet, accessible only to those who traveled and stayed there, the esoteric teachings of an introverted community following its ancient spiritual practices over centuries. After the expulsion, the understanding grew that it was the responsibility of the Tibetan diaspora to make their tradition available to the world, that it was for this that they had had to leave their homes. Fundamental to the Buddhist expression of compassion is the image of the boddhisattva, the one who on reaching enlightenment chooses to continue to serve, helping other sentient beings along the way.

This particular group, led by Tara Tulka Rimpoche, came from a tradition of ritual chanting in which each monk, entering the monastery in early childhood, learned to chant several notes on the same breath, a chord. They had begun to offer their chanting rituals before Western audiences, but only in spaces sanctified to prayer and under conditions of careful solemnity. This was what they had come to do in Iran, I learned. Because the ecumenical impulse is minimal in official Shiite Islam, they were to present their rituals in Christian churches. Muslims regard Jews and Christians as People of the Book who have accepted some portion of the revelation of a prophet of the One God and thus are on the road to the truth; but Buddhists are regarded as pagan idolaters.

The agenda of the monks, then, was religious: to present their rituals in such a way that a spiritual meaning and benefit would be imparted to their audiences, bringing them some fractional step closer to enlightenment. But the agenda of their hosts was very different. The monks—along with other groups of actors, dancers, and singers from all over the world—had been brought to Iran as an "act" for the Shiraz Arts Festival. The shahbanu, the empress, had created the Shiraz Festival, with performances ranging from the exotic to the avant-garde, scandalizing traditional Iranians—postmodern experimentation in provincial Shiraz, city of poets and gardens, simulated rape, incitement to blasphemy. The festival came to represent the efforts of the monarchy to introduce dangerous and alien innovations. Two years later the religious community, protesting the tyranny of the shah, demanded free speech and, in the same ultimatum, demanded the right to censor the festival.

In Athens, speaking with and through the interpreter, I learned that the monks knew almost nothing about Islam and were anxious to learn, but that no arrangements had been made for them to do so. So I offered to take them the next day to some historic mosques where tourists were permitted, and to the lodge of a Sufi teacher.

But first I would attend their ceremony, scheduled to take place in the evening in a little Presbyterian mission church in downtown Tehran. A sort of island of Tibetan Buddhism had been created in that church: in front of the tall-backed chairs and choir stalls, scarlet and orange cushions had been set out, a low table covered with brocade, long Tibetan horns, bells, images. Above them, a portrait of the Dalai Lama partly obscured a cross outlined in neon.

It was summer and very hot. Tehran is built on a mountainside, the wealthiest houses on the cooler slopes, and as you go, literally, downtown, the city gets hotter, poorer, and more crowded. Open watercourses run alongside the streets; although these are by no means the open sewers that Westerners tend to see in them, there is a vast difference between the crystalline clear fountains of the Niavaran Palace at the top of the city and the dirty water that reaches the bottom. Smog is heavy, and the traffic is horrendous, old, fuming vehicles and aggressive drivers. So I was not popular with my husband when I told him to meet me at the mission church. The place was packed, but I had saved him a seat. He arrived caustic and complaining, and matters got steadily worse as we waited for the monks to begin. And waited. Lights for filming were set up at the sides of the sanctuary. The little church got hotter, Barkev became more sarcastic about my eccentric tastes. And still we waited.

Later I learned that the monks had been outside the whole time, sitting in a bus and declining to begin, because they had discovered that they were to be taped for television. Their sense of mission, their sense of the efficacy of their ritual for those who heard it, required actual presence, and they refused to have it canned. The Arts Festival officials were furious. "Tell them they have to perform," they said to the interpreter. "No," he said, "they don't. Telling them that won't do any good at all." Officials threatened cancellation, lawsuits, expulsion from the country, jail. The monks sat serenely in their bus. The waiting audience became increasingly restive. One hour after the ceremony was supposed to begin, technicians came in to dismantle the TV lights, and finally the monks entered.

They walked with dignity to the front of the church and sat in a little circle of Tibetan culture surrounded by evangelical Protestantism, like Victorian imperialists dressing for dinner in the jungle. They began to chant. Suddenly that little church was cool and calm; dirt and sweat and anger were gone. Barkev said afterward that he had felt the temperature had gone down ten degrees, his pulse slowed. Years later, when Vanni went on a whale-watch sail on a little boat full of pushy, impatient people, she described the way their fractiousness simply disappeared as soon as they saw and heard the whales, and Barkev compared those moments with the first time he walked in virgin forest in Mindanao, just before the loggers arrived.

The next day when I came to the hotel to pick the monks up, I found that although they had agreed to allow an audiotape to be made the night before, for "archival purposes," instead of a video, they were indeed being expelled from the country and their contract canceled. We had only a few hours before their flight, but they were unperturbed and still determined to come with me to meet the Sufi teacher, Dr. Nurbakhsh.

Sufis are the mystics of Islam, the seekers of religious experience rather than the punctilious observance of rules, the lovers of paradox and play. The tension in which they live with orthodoxy replicates itself in many traditions, where a few go beyond or beneath the belief of the many, rejecting, reinterpreting, and seeking, often subject to persecution. I had once asked Dr. Nurbakhsh if he could teach me to be a Sufi, and he had responded that although a Christian could become a Sufi, he himself could not guide a Christian to Sufism because he only knew how to guide people through— and by implication beyond—Islam.

Sufis use stories, humor, and poetry in their teaching, so I had invited a friend whose Persian was better and more literary than my own to come and translate. A religious dialogue took place between the Sufi teacher and the Tibetan abbot, sitting on Persian carpets and sipping tea, with the monks around them in a circle. The Sufi was small and wiry, with a dramatic mustache, flashing eyes, and staccato movements—an imp, a maker of divine mischief. The monks were big and serene and smooth muscled. They sat cross-legged, leaning forward, following each speaker carefully. First the abbot spoke, with measured courtesy, in Tibetan. Then the first translator translated into English. Then the second translator translated the English into Persian. Then the Sufi answered, sparks jumping. Back the translations went, the eyes of the monks going from one to the other as if they were watching a volleyball game.

"What do you do here?" asked the abbot. Translator followed translator.

"We try to become nothing," said the Sufi.

"Is that the nothingness which is death?" said the abbot.

"The wave rises and falls and loses its separateness in the sea." Back and forth.

After scarcely a dozen exchanges, Nurbakhsh quoted a line of poetry, translated as "the fire of love did not burn so brightly in the beginning, but was fanned by those who passed by."

The abbot reflected for long moments. Then he said, "We have that too. We call it boddhisattva."

He looked at his watch and rose, satisfied. All the monks rose, gathering their robes, thanks were said, and I walked out and stood with them as they boarded their bus. There in a two-day period I had examples of total noncomprehension between cultures and of recognition that transcended style and place and language. As the bus drove away, I stood on the street corner, waving good-bye to that small group with whom I had no word in common, and I discovered that tears were running down my cheeks.

The singing of whales has become for many people a paradigmatic experience of the sacred, an encounter with another species living in a totally different medium, suddenly known as kin rather than as stranger. The moment of recognition is a moment of self-knowledge as well. This is what I imagine happening to the first Christians after the execution of Jesus: bewildered and bereaved, their beloved teacher gone, they suddenly began to recognize him in the faces of strangers and in the very bread they ate. The Gospel stories do not tell us that in those days they looked at the sky and the grass as well, saying, Look, he is here, or that they recognized him in birds or insects. If they had, the history that followed might have been very different, but they were Jews, their visions of human interactions with the divine far removed from the insights of pantheism.

St. John's Gospel tells how Mary Magdalene went to the tomb on the second day after the crucifixion. Finding it open and empty, she turned in distress to a stranger, taking him to be the gardener, and asked where the body had been taken. "Mary," he said. And in that moment of being recognized she recognized him as well. But ecclesiastical tradition talks of bodily resurrection instead of looking with new eyes at every gardener, traveler, stranger. In spite of the stories of men and women entertaining angels unawares, we remain stubbornly blind to angel faces—all the faces—in the crowd, and angel voices singing in the forest or the sea. The entrepreneurs of the Shiraz festival hardly saw the human faces of that group of monks as they ordered them to perform.

I had picked the Tibetans up in the transit lounge out of the habit of curiosity and because I had once heard a Tibetan Buddhist sermon that stayed in my memory. Curiosity is a good place to start if one is going to encounter the sacred. One of the great tragedies in Western history has been the notion that the inquiring mind, questioning orthodoxy, is therefore irreligious, so that only once in a while do very distinguished and elderly scientists admit that behind their curiosity there is an attitude of wonder. I worry sometimes about the children of religious parents whose piety is protected by the cultivation of large areas of ignorance and the suppression of imagination, even as their virtue is protected by a deliberate concealment of the realities and delights of the human body. The habit of curiosity can also be corrupted, turned into a restless search for novelty, which is equally blinding.

In our overly busy culture, only a few experiences—sailing alongside singing whales in the advancing dusk, listening to the chanted harmonies of Tibetan refugees, looking at the enigmatic smiles of archaic Greek statues—cut through to the response of wonder, and not for everyone. A recognition of the wonder of the natural world can transcend human concerns, even the concerns of hunger or fear or impending death, but no such recognition distracted the teams felling the great trees in Mindanao for sale to Japan. The alchemy of recognition does not always work.

Going alone into primeval forest is moving not because

the forest is big or ancient or even because it separates the visitor from routine concerns, though all these things are true, but because it is orderly, the working out of patterns of ecological interdependence and coevolution more complex than any symphony. The attitude of mind that responds to the forest is the habit of searching for and responding to pattern, a consciousness ready to be schooled by complexity. The size of a forest or even of a whale cannot be central, for the analogs of great forests can be seen in ponds, even transiently in tide pools, more visible and comprehensible to children and newcomers than the great forests or prairies. More and more, I believe, we will learn to look for epiphanies by looking through microscopes. The challenge for parents and educators is to create the readiness to respond, the quality of attention that makes recognition possible: pattern matched with pattern, vagrant awareness welcomed, empathy established. The kind of subtle interior shift that happened to me as I toured around Tehran with the Tibetans is only described in love songs.

What was most moving to me was the agility with which the Buddhist and the Sufi moved from courtesy to mutual recognition. It used to be that explorers or traders or missionaries looked at the exotic peoples they encountered as orthodox mullahs would look at Tibetan monks, as benighted, but you cannot learn easily or at any depth from those you look down on. I have spent my life as an anthropologist trying to learn how to share enough with strangers to make learning possible, learning to identify divergent premises instead of taking my own for granted, and to accept a broader or more ambiguous view than common sense. The basic challenge we face today in an interdependent world is to disconnect the notion of difference from the notion of superiority, to turn the unfamiliar into a resource rather than a threat. We know we can live with difference—men and women for instance have lived together throughout history. We know we can benefit from difference. But the old equation of difference with inferiority keeps coming back, as fatal to the effort to work together to solve the world's problems as the idea of competing for a limited good.

Today encounters between different spiritual and artistic

traditions happen more and more often, in circumstances
that do not label one side as dominant or superior, and urban
children have heard rumors of multiple forms of traditional
wisdom. They will need a new kind of openness and responsiveness to discover all they must learn along their way.

For when great changes and revolutions occur, whether
they are social or technological, they must be lived out by
men and women who matured under the old order. We know
that the people of the Soviet Union were not educated for
democracy, so we must hold on to the hope that somewhere
in their growing up they encountered themes that will help
them in devising new forms of civility as adults. Similarly, we
know that the black and white populations of South Africa
have not been educated to regard each other as fellow citi-
zens, so we must pin our hopes on their capacity to transfer
patterns from one context to another. Even when hate
becomes obsolete, it is harder to adapt without habits of
questioning and independent observation, just as it is hard to
learn any new skill when existing skills are based on a rote
learning of technique. It may be that both kinds of learning
occur only in the context of a certain degree of self-
confidence and self-respect, and must be carried by some
affirmation of continuity. We know that today's children are
being educated by parents who have not yet learned to live
lightly on this earth. Fifty or even twenty years from now,
they may find themselves living differently, feeling either
harassed and deprived by new circumstances, or perhaps
experiencing an unpredicted homecoming and sense of
peace.

Beyond either relativism or the search for absolutes,
learning can be practiced as a form of spirituality through a
lifetime. We started from participant observation and the
necessity for improvisation, asserting the need to act and
interact with others without complete understanding, learning along the way, and we argued that improvisation can be
both creative and responsible. We have explored ways of
embracing myths and metaphors and multiple layers of
truths, education through lessons that are different at every
encounter. The self is constructed from continuing uncer-

tainty, but it can include or reflect a community or even the entire biosphere, can be both fluid and stable, can be fulfilled in learning rather than in control.

Again and again we have rejected the "rhetoric of merely," the rhetoric that treats as trivial whatever is recognized as a product of interacting human minds that may then go on to some other product or point of view. Because it is not possible to stand aside from participation until we know what we are doing, it is essential to find styles of acting that accept ambiguity and allow for learning along the way. Perception, attention, grace, all of these, varied or sustained, provide materials for constructing both self and world, and patterns for joining in the dance.

In Washington, DC, the Smithsonian Institution has an annual folk festival to which it brings Balinese dancers, shamans from the Orinoco, Tlingit storytellers, gospel singers from the Deep South, performing outdoors on the Mall. I have heard that they all stay in the same hotel, and I love to imagine the tentative explorations and conversations, the dawning recognitions of collegiality. We have that too. We call it boddhisattva.

Acknowledgments
and Sources

THIS BOOK WAS MADE POSSIBLE above all by my experiences living in other cultures and by the hospitality and tolerance shown to me during these sojourns. Part of the pleasure of the writing has been revisiting notes and memories going back many years, bringing them out and considering the ways in which they illuminate the questions I am asking today. I was in Israel as a high school student in 1956–57, visited in 1988 after a thirty-year absence, and returned for six months in 1989 as a researcher. I was in the Philippines with my husband, Barkev, between 1966 and 1968 as a young professor and field-worker. We lived in Iran before the revolution, from 1972 to 1979, with our daughter, Vanni, teaching, researching, and planning for new and developing educational institutions. Each of the memories I have picked up along the way represents an encounter with a place and a time that goes beyond the particular but might have been invisible without the contrasts of strangeness, for one is forced by cultural difference to question assumptions and struggle for active understanding.

I am grateful to friends and acquaintances in each of the societies discussed here, who offered me their hospitality and the opportunity to try to understand their lives. It would not be possible or appropriate to list them all, and some appear only under disguised first names, but there are a few key people, some of them no longer living, and institutions that must be thanked explicitly.

In Israel, I wish to express my gratitude to Phyllis Palgi and her husband, Yoel; Mordechai Kamarat; Joyce and Louis Miller; Ran Avrahami; and Aharon Shavit. Among institutions I am especially grateful to the Hebrew University Secondary School, where I was a student on my first trip; the Barcai Institute for Family and Marital Therapy, which arranged my first return; and the Van Leer Jerusalem Institute, where I was a guest in 1989. Some of my work in Israel was funded by grants from the Lucius N. Littauer Foundation and the Foundation for Mental Research to the Institute for Intercultural Studies.

In the Philippines I was associated with the Anthropology Department of the Ateneo de Manila University and the Institute for Philippine Culture directed by Frank Lynch, S.J., a scholar of Philippine values. My husband was associated there with the Harvard Advisory Group Interuniversity Program on Management Education, supported by the Ford Foundation. I also want to express gratitude to Jaime and Maribel Ongpin.

In all discussions of Iranian culture, I am indebted to other members of a national character study group that met with me and Barkev for several years in the seventies, especially J. W. Clinton, H. Safavi, M. Soraya, and M. Tehranian. Some of these ideas have also been developed in discussions with other anthropologists working in Iran, especially William Beeman. In Iran I was associated with Damavand College and Reza Shah Kabir University, now no longer in existence. My husband taught at the Iran Center for Management Studies, and I am grateful to many of his colleagues there, especially Habib Ladjevardi. My initial work in Iran was supported by the Wenner-Gren Foundation. St. Paul's Church in Tehran published my booklet on cross-cultural misunderstandings, for which I drew on discussions with members.

Each of these experiences in faraway places was followed by a period of considering American society with new eyes, realizing that the same lessons were available here, for a rapidly changing society embracing many cultures and traditions—multiple systems of meaning—is a school of life. Since the Iranian revolution in 1979, I have been writing primarily about the United States. In the United States I have had the benefit of participating in several communities of thought and conversation, including the annual conferences of the Lindisfarne Fellowship, founded by William Irwin Thompson. I have found it especially useful when face-to-face meetings were supplemented by the virtual communities of computer conferencing, as they were in the Learning Conferences sponsored by Shell, Volvo, and AT&T and in the Global Business Network, established by Peter Schwartz and Stewart Brand.

The intellectual roots of this book are in anthropology, and although I provide references below for a few specific works referred to, the habits of thought and observation of cultural anthropology are largely taken for granted. Every reader comes from a somewhat different background, however, and some of the familiar furniture of my own mind may be unfamiliar to others. This is particularly likely to be the case with two intellectual pioneers whose thinking I encountered so early in life that it has largely meshed with my own. Thus, I want to acknowledge here the contributions of my parents, Margaret Mead and Gregory Bateson, whose ideas may appear sea-changed, having been synthesized over time with other ideas and influences. I have written elsewhere about both of them and nursed their intellectual legacies. Here I simply speak from what has become a part of me.

This book owes a part of its genesis to reactions to my previous book. I knew when I wrote *Composing a Life* that most of its message was applicable to men as well as women. However, our society tends to regard women as exceptions to the full human condition and men as representative, so it was somewhat mischievous of me to write a book about human beings in the form of a book about women, and some men have complained that they gained valuable insights about themselves from the book but would not have read it on their own. Today we can insist that research on such topics as drug reactions be conducted with samples including females as well as males so that we can distinguish differences from similarities, so I felt it was important that I attempt to make the resources offered by my work available to men. I want to thank those readers who urged me in that direction.

It was not until after the appearance of *Composing a Life,* when I began to be invited to speak to groups of educators and faced their questions, that I fully realized that I had written, unlabeled, a book about learning from experience. I am in debt to those who asked questions I had not yet begun to think about, for they made me aware of the ferment in the field of education and set the course of this new book.

A central theme of this book was first stated in a paper given just before I went to Iran at the annual meeting of the American Anthropological Association in 1971 on conversations involving participants with disparate codes, later published as "Linguistic Models in the Study of Joint Performances," in *Linguistics and Anthropology in Honor of C. F. Voegelin,* M. Dale Kinkade, K. L. Hale, and Oswald Werner, eds. (Lisse: Peter de Ridder Press, 1975), pp. 53–66. An account of my research with Margaret Bullowa's project on child language learning at MIT was published in "Mother-Infant Exchanges: The Epigenesis of Conversational Interaction," in *Developmental Psycholinguistics and Communication Disorders,* Doris Aaronson and R. W. Rieber, eds., *Annals of the New York Academy of Sciences,* vol. 263 (1975), pp. 101–113.

Several chapters of this book are derived from lectures of mine that have circulated in taped form or from previously published essays, heavily revised and reshaped. Chapter 1 is partly based on a keynote address given in 1992 to the Western States Communication Association Convention and published in transcript form as "Joint Performance Across Cultures" in *Text and Performance Quarterly,* vol. 13, no. 2 (April 1993), pp. 113–121. Chapter 2 is based in part on my essay "Insight in a Bicultural Context," *Philippine Studies,* vol. 16, no. 4 (1968), pp. 605–621, and a further portion of the same essay appears at the end of Chapter 10. Chapter 6 was published in an earlier version as "The Construction of Continuity" in Suresh Srivastva and Ronald E. Fry, eds., *Executive and Organizational Continuity: Managing the Paradoxes of Stability and Change* (San Francisco: Jossey-Bass, 1992). Chapter 8 includes portions of an unpublished paper, "Notes on the Problems of Boredom and Sincerity," distributed to participants in Burg Wartenstein Symposium No. 59, *Ritual: Reconciliation in Change*, July 21–29, 1973. Chapter 9 combines a lecture given to the Isthmus Institute in Texas on the Gaia hypothesis in 1990 and a keynote address given at the American Association of Colleges for Teacher Education in the same year. Chapter 11 incorporates a paper given at an American Anthropological Association Symposium on multiculturalism at the association's annual meetings in November 1992. Many of the vignettes about differences between Iranian and American cultures that occur throughout the text were first presented in a booklet called *At Home in Iran* (Tehran: St. Paul's Church, 1973 and 1976).

Any author must make decisions about how often to supply references. On the one hand, full references are valuable to protect the rights of other writers and researchers and to provide resources for those who follow similar paths. On the other hand, an excess of scholarly machinery creates a block between writer and reader. My own policy is to provide footnotes only for exact quotes and for information that I myself have obtained from a single, specifiable source. The same philosophy has led me to use spellings of Persian or Hebrew words that I believe will offer comfort to the reader rather than erudition. I have felt it necessary to include from time to time examples from a kind of society I have never myself studied, and these have been drawn primarily from the superb body of ethnographic work on the !Kung San of the Kalahari. Among the more accessible books on the San are:

Richard B. Lee, *The Dobe Ju/'hoansi* (Fort Worth, Tex.: Harcourt Brace, 1993). A first edition in 1984 was titled *The Dobe !Kung.*

Elizabeth Marshall Thomas, *The Harmless People* (New York: Alfred A. Knopf, 1958).

Marjorie Shostak, *Nisa: The Life and Words of a !Kung Woman* (Cambridge, Mass.: Harvard University Press, 1981).

*   *   *

In a number of cases I evoke a body of work simply by the name of the author. Fuller references appear here:

William Beeman, *Language, Status, and Power in Iran* (Bloomington, Ind.: Indiana University Press, 1986).

Margaret Bullowa, *Before Speech* (New York: Cambridge University Press, 1979).

Karl W. Deutsch, *The Nerves of Government,* 2nd ed. (Cambridge, Mass.: MIT Press, 1966).

Robert Edgerton, *Mental Retardation* (Cambridge, Mass.: Harvard University Press, 1979).

Erik H. Erikson, *Toys and Reasons: Stages in the Ritualization of Experience* (New York: W. W. Norton, 1977).

George Foster, "The Idea of Limited Good," in *Peasant Society: A Reader,* ed. by Jack M. Potter (Boston: Little, Brown, 1967).

E. D. Hirsch, *Cultural Literacy: What Every American Needs to Know* (Boston: Houghton-Mifflin, 1987).

H. E. Huntley, *The Divine Proportion: A Study in Mathematical Beauty* (New York: Dover, 1970).

Theodora Kroeber, *Ishi in Two Worlds: A Biography of the Last Wild Indian in North America* (Berkeley: University of California Press, 1961).

Jean Lave and Etienne Wenger, *Situated Learning: Legitimate Peripheral Participation* (New York: Cambridge University Press, 1991).

Alan Lomax, *Folk Song Style and Culture* (New Brunswick, N.J.: Transaction Books, 1978).

Konrad Lorenz, *King Solomon's Ring: New Light on Animal Ways* (New York: T. Y. Crowell, 1952).

James Lovelock, *Gaia: A New Look at Life on Earth* (New York: Oxford University Press, 1987).

Frank Lynch and Alfonso de Guzman, *Four Readings on Philippine Values* (Quezon City: Ateneo de Manila University Press, 1970).

John von Neumann and Oskar Morgenstern, *The Theory of Games and Economic Behavior* (New York: John Wiley, 1964).

Paul Radin, *Crashing Thunder: The Autobiography of a Winnebago Indian* (1926) (Lincoln: University of Nebraska Press, 1983).

Oliver Sacks, *A Leg to Stand On* (New York: Harper-Collins, 1984).

Daniel B. Schirmer and Stephen Rosskamm Shalom, eds., *The Philippines Reader: A History of Colonialism, Dictatorship, and Resistance* (Boston: South End Press, 1987).

Carol Stack, *All Our Kin: Strategies for Survival in a Black Community* (New York: Harper & Row, 1974).

Deborah Tannen, *You Just Don't Understand: Women and Men in Conversation* (New York: William Morrow, 1990).

Howard Thurman, *The Inward Journey* (Richmond, Ind.: Friends United Press, 1971).

Malcolm X, *The Autobiography of Malcolm X*, ed. by Alex Haley (New York: Grove Press, 1965).

I worked on this book on and off over a three-year period with flexibility made possible by George Mason University, which supported me in travel and manuscript preparation, and I spent a critical two months at the MacDowell Colony in Peterborough, New Hampshire, a setting beyond compare for intensive work and a fascinating example of a transient community. My agent, John Brockman, and my editor, Susan Moldow, have done splendidly in supporting me in the transition from aspiration to reality. Others who have read and commented on portions of this manuscript include Diana Feige, Mary Garland, Richard Goldsby, Barbara Kreiger, Alan Lelchuk, Harold Morowitz, Frank Ortega, Roy Rappaport, Lisa Raskin, and David Sofield.

My husband, Barkev, and my daughter, Vanni, have grown accustomed to the fact that I use what I learn with them and from them in my work. Barkev has always read and commented on multiple drafts, and Vanni too has taken up that role. Her voice will bring this book to life on tape. My thanks to Barkev and Vanni for support, caring, patience—all those things—but above all for the pleasures of life entwined with active minds.